# Seed Processing

# Seed Processing

## A Practical Approach


**Rakesh C. Mathad**
**A.S. Channaveeraswamy**
**B.S. Vyakaranahal**
**V.K. Deshpande**



**NEW INDIA PUBLISHING AGENCY**
New Delhi – 110 034

**NEW INDIA PUBLISHING AGENCY**
101, Vikas Surya Plaza, CU Block, LSC Market
Pitam Pura, New Delhi 110 034, India
Phone: + 91 (11) 27 34 17 17 Fax: + 91 (11) 27 34 16 16
Email: info@nipabooks.com
Web: www.nipabooks.com

Feedback at feedbacks@nipabooks.com



ISBN: 978-93-85516-08-5



Composed and Designed by NIPA

*Dedicated to*

**Dr. B S VYAKARANAHAL**

Professor (Seed Science and Technology)

*Our beloved teacher...*

# Preface

*"Don't judge each day by the harvest you reap but by the seeds that you plant."*

— **Robert Louis Stevenson**

In order to understand the importance of quality seed, it is very important to know its scientific basis in terms of processing, seed treatment and packaging. We wrote this book to provide a framework for learning these basic shop floor activities to make 'raw seed' in to 'performing seed'. Seed processing will contribute to bumper harvests and without this timely up-gradation even superior varieties also cannot deliver objectives for which they are released. Seed processing directly related to its performance in the field and positive socio-economic attributes. It improves its physical purity, germination, vigour and health status. To understand seed processing it is important to study the attributes practically and students can get a first hand understanding of different processes involved in up-gradation with effect on quality attributes.

In this book a basic understanding on seed processing with an activity intend to find the effect of each process on seed quality is written. Apart from basic seed processing terminology and shop floor assembly of equipments, emphasis is given on different stages of seed processing like, extraction, up-gradation, blending, drying, treatment, fumigation, packaging and storage. Also students can find effect of these processes on seed quality like mechanical injury after processing, standardization of screen sizes for different crops, importance of seed moisture and suitability of certain crops for storage. This book contains effect of seed quality enhancements like pre-sowing treatment, pelleting and priming on quality. This book highlights certain problems in seed processing like post harvest losses during processing, processing efficiency, 'stackburn' and safety at shop floor with issues like personal protection and ESH (environment, safety and health).

In each of thirty two practical sessions with brief introduction contain a detailed activity and tools to collect data. There are worksheets and methods explained for the students who conduct these experiments during practical sessions. This book can be used as practical manual for undergraduate and post graduate courses in seed processing.

We are very receptive to suggestions and constructive criticism to improve this book further. We lived the authors dream, thanks to Sumitji, team at NIPA and their decision to publish this book.

**Rakesh C. Mathad**
**A.S. Channaveeraswamy**
**B.S. Vyakaranahal**
**V. K. Deshpande**

# Contents

# Seed Processing Terminology

## A

**After-ripening.** The physiological maturation processes which occur in e.g. seeds and fruits after harvest or abscission. After-ripening is often necessary for immature seeds to become germinable. Also used for the seed handling process itself.

**Application Rate:** The application rate of Seed Treatments, Seed Treatment Products and/or Seed Treatment Components. For liquid Seed Treatment Products, rates are typically expressed in ml or in mg ai/kernel on the label.

**Aspirator or blower**: A processing equipment in which when a seed mixture is introduced in to a upward moving air stream, light particles with a low terminal velocity are carried along with the flow where as heavy seed falls through.

## B

**Batch:** A defined quantity of seed material, packaging material or product processed in one process or series of processes so that it could be expected to be homogeneous.

**Bio-priming** : is a process of biological seed treatment that refers combination of seed hydration (physiological aspect of disease control) and inoculation (biological aspect of disease control) of seed with beneficial organism to protect seed. It is an ecological approach using selected fungal antagonists against the soil and seed-borne pathogens. Biological seed treatments may provide an alternative to chemical control.

**Batch number (or lot number):** A distinctive combination of numbers and/or letters which specifically identifies a batch.

**Bulk Seed**: Any seed lot or batch which has completed all seed processing stages up to, but not including, final packaging.

**Blending:** It is simply combining two or more seed lots of same variety or hybrid of specified quality. The blending may involve lots which may have different quality in terms of germination but the blended lot meet the minimum standards.

**Bucket Elevator**: Conveying equipment used in processing to more seed to overhead storage bins, designed to cause minimal mechanical damage to seed.

## C

**Calibration:** The adjustment of seed treatment equipment to apply the target rate of slurry or Seed Treatment Product and the verification thereof.

**Chaff:** Thin dry bracts or scales, especially the dry bracts enclosing mature grains of cereal grasses, primarily removed during threshing but which may be present in low levels in commercial seed. More generally, seed debris.

**Class** (of seed and seed crop) – Refers to the generations of pedigreed seed and seed crops, such as Breeder, Foundation, Certified and truthfully labelled seed which have met the standards prescribed by recognized seed and seed crop certification agencies.

**Cleaning, pre-(seed):** Separation of seed from other species and non-seed fragments such as fruit fragments, leaves or stems. Cleaning may be undertaken by sifting, blowing, winnowing, flotation etc.

**Colour Sorter**: A processing equipment which removes defective individual seeds one at a time based on differences in reflected colour or brightness under one or two selected visible or infrared light wavelengths.

**Conditioning of seed** – A term used to describe the cleaning of seed, usually to improve mechanical purity.

**Cryopreservation or cryo-storage.** Maintaining tissues or seeds for the purpose of long time storage at ultralow temperature, typically between -150°C and -190°C. In normal cryopreservation the sample is pre-treated with a cryo-protective substance, followed by slow, controlled freezing. This method is often employed for recalcitrant seeds.

## D

**Dis-infection, seed:** It refers to eradication of fungal spores present within the seed coat or more deep seated tissues. For effective control the fungicide must penetrate into the seed to kill the fungus.

**Dis-infestations, seed:** It refers to the destruction of surface borne organisms that contaminated the seed surface but not infected the seed. Chemical dips, soaks, sanitation, fungicides applied as dust, slurry or liquids have been found successful.

**Desiccant:** Chemical compound that has a high moisture absorption affinity and can be used for desiccation or maintaining a low humidity when stored together with e.g. seeds. Common desiccants are SiO, CaO.

**Dry seeds**: This type of crops like sponge gourd, ridge gourd, melons, onion, radish, carrot, spinach or most of the exotic vegetables. Here the seed is not attached to flesh at the time of harvesting. The flesh holding the seed will be either completely dried or seed bearing on the dry umbels like in onion

## E

**Enhancement, seed:** Seed treatment to improve lustre, quality and performance like polymer coating or priming. Also known as functional seed treatment.

**Equilibrium moisture content (EMC):** Moisture content of seeds in equilibrium with atmospheric humidity at a given temperature. The EMC is influenced by hygroscopic character of the seed storage material viz. low for oily seeds and high for seeds rich in protein and carbohydrates

**Equilibrium Relative Humidity (ERH)**: The percentage relative humidity at which a given EMC is expressed at a specified temperature.

**Encrusting, seed**: encrusting applies to less material so the original seed shape is still visible.

## F

**Fluming or flotation:** It is a method of removal of light weight and immature seed by floatation in water as part of seed extraction activity.

**Film coating**: Seed treatment with materials which regulate seed water availability and gas exchange, there by controlling timing of germ and emergence.

## G

**Genetic purity** – Trueness to type or variety.

**Grading**: Separation of seed crop produce in to different categories according to size, weight, colour and quality.

## H

**Handling seed:** Handling includes the movement or products and treated seed, including but not limited to loading, unloading, weighing, bagging, sewing, stacking, and planter loading and operation.

**Hardening, seed**: It is a process of seedling grown under green house or shade net before transferring to harsher field conditions by reducing watering, shelter and fertilizer & so on.

**Hazardous components:** Components which present health, safety or environmental hazards.

**Hermitic storage**: Storage inside air-tight containers. This is very essential for maintaining low moisture content required for successful long term storage of dried orthodox seeds.

**Hygroscopic:** Sensitive to moisture, absorbing or losing water. Pores on seed coat open and close according to humidity because they contain hygroscopic tissue.

## I

**Imbibition:** The process of initial water uptake by seeds prior to germination. Imbibition is an entirely physical process and also non-viable seeds imbibe.

## L

**Labels** – issued to identify the crop, variety, standards and class of a seed lot. Sometimes called "tags."

**Liquid Density Seperation**: Method for cleaning seeds from particles with higher or lower specific density by submerging in water or other liquid. Used e.g. for separation of empty and filled seed and mechanically damaged seed

**Longevity, seed**: The period of time seed will maintain viability in storage under a given set of storage conditions. Often used equivalent to storability

## M

**Material Safety Data Sheets (MSDS):** The MSDS is a detailed informational document prepared by the manufacturer or importer of a hazardous chemical. It describes the physical and chemical properties of the product, and is a tool for communicating safe handling and environmental protection requirements for chemical products. For pesticide products such as Seed Treatment Products, the information on the label takes precedence over the information on the MSDS.

**Moisture meter:** Instrument for quick measurement of moisture content of seed without drying. The instrument measures the electrical properties of the seed tissue, which is correlated with the content of water (moisture content). Moisture meters must be calibrated for each species with standard method of moisture content measurement i.e. oven-drying.

## N

**Noxious weed** – A weed or plant that is considered undesirable and so categorized by Seed Act or Rules or Indian Minimum Seed Standards

## O

**Orthodox:** Term used to describe seeds which can be dried down to a low moisture content of around 5% and successfully stored at low or sub-freezing temperatures for long periods. There is practically no metabolism in dry, cooled orthodox seeds, but they may deteriorate by general ageing and thus ultimately lose their viability

**Other crop seed** – One of the four components of a seed purity test and usually refers to the number of seeds of other crop kinds in the seed sample being tested.

## P

**Packaging:** The container holding the untreated or treated seed or holding Seed-normally a aluminium foil or a tin

**Personal Protective Equipment (PPE):** Equipment that is worn by employees to mitigate hazards of a process. For seed treating operations, PPE is typically means to reduce exposure of operators to seed treatments and treated seed dust. Such PPE includes but is not limited to long-sleeved shirts; long pants; shoes; socks; goggles; chemical resistant gloves; and respirators.

**Pelleting:** Procedure by which individual seeds are provided with an envelope of adhesive material containing e.g. nutrients, microsymbiont inoculant and/or pesticides. In addition to providing these beneficial compounds, pelleting facilitates mechanical sowing because of the more uniform seed size

**Physical purity** – Refers to the degree of freedom of a seed lot from seeds of other crop kinds, weed seeds and inert matter.

**Physiological maturity:** General term for the stage in the life cycle of a seed when development is complete and the necessary biochemical components for all physiological processes are active or ready to be activated.

**Priming:** Pre-treatment method to promote rapid and uniform germination. The seeds are soaked in a liquid solution (e.g. polyethylene glycol (PEG), sugar or salt) of sufficiently low water potential to regulate moisture content at a level where the germination process initiates but radicle protrusion is prevented.

**Processing, seed**: Seed handling methods from collection to storage, usually a collective term

applied to extraction, cleaning and drying.

**Protection, seed**: To protect the seed and young seedling from organisms in the soil which might otherwise cause delay of the seed before germination.

**Purity:** Proportion of clean, intact seed (according to pure seed definition) of the designated species in a seed lot, usually expressed as a percentage by weight.

## R

**Recalcitrant:** Term used to describe seeds that cannot survive drying below relatively high moisture content (30-40%) and, for tropical species, do not tolerate low temperature. The seeds rapidly lose their viability and cannot be successfully stored for long periods

**Relative humidity (RH):** The actual amount of water vapour in the atmosphere as a percentage of that contained in an atmosphere saturated with water at the same temperature.

## S

**Sampling** – The method by which a representative sample is taken from a seed lot to be used for analysis.

**Sample:** In the context of seed testing, a small representative quantity drawn from a seed lot. The different types of samples in seed testing are primary sample, composite sample, submitted sample and working sample.

**Seed Flow:** The uniformity and freedom of flow of seed through a system, generally through a seed conditioning or treating plant; or through a planter. Poor seed flow may be slow or inconsistent seed flow, or plugging of auguer or conveyors or other handling equipment. Seed treatments may positively or negatively impact seed flow.

**Seed lot**: A specified quantity of seed of the same species, provenance, date of collection and handling history, and which is identified by a single number in the seed documentation system.

**Seed lot system**: A seed lot system is a system according to which successive batches of a product are derived from the same master seed lot at a given passage level. For routine production, a working seed lot is prepared from the master seed lot. The final product is derived from the working seed lot and has not undergone more passages from the master seed lot than the vaccine shown in clinical studies to be satisfactory with respect to safety and efficacy. The origin and the passage history of the master seed lot and the working seed lot are recorded.

**Seed Moisture:** The free water content of seeds, typically measured as a percentage.

**Seed Production:** The process of growing crops to be sold as seed, instead of growing crops to sell as food or feed. Because each seed will become a crop plant, seed production requires high standards for quality and germination

**Seed Treater:** Equipment designed to apply seed treatments to seed. Such equipment should be designed so it can be calibrated to accurately and uniformly apply the product to seed. Numerous types of seed treaters exist.

**Seed Treatment:** Seed treatment is the application of biological organisms and chemical ingredients to seed to suppress, control, or repel plant pathogens, insects, or other pests that attack seeds, seedlings or plants. Seed applied technologies such as inoculants, herbicide safeners, micronutrients, plant growth regulators, seed coatings, colorants, etc. may also be applied to the seed. Treated seed is intended for planting only and not for food or feed uses.

**Seed Treatment Polymers:** Products added to seed treatments whose primary function is to reduce dust of treated seed and to improve retention of seed treatment active ingredients on the seed

**Screens**: Screens with different-sized openings are used to separate seeds from chaff and distinguish seeds. The screen number denotes the number of openings that will cover a one inch line. A screen is selected with openings just large enough to let seeds drop through without the chaff or as in the case of larger seeds, a screen selected to allow the chaff to drop through without the seeds.

**Slurry:** The combined treating composition for application to seed. It may be as simple as a single ready to use product, or a combination of several Seed Treatment Components and water.

**Sorter**: Pneumatic sorting tables are used to separate seeds by weight, and a grain cleaning column is used for cleaning seeds from other substances with different aerodynamic properties. Gravity seed cleaners and electromagnetic seed cleaning machines sort the seeds by shape and surface texture.

## T

**Threshing:** Disintegration and extraction of seeds from dry fruits by mechanical impact to the fruit, e.g. flailing, beating, trampling, stamping, or by threshing machines.

**Treated seed:** Seed that has been treated with a "Seed Treatment Product" like fungicide, pesticide, polymer or any seed quality enhancing material.

## U

**Upgrading:** Improving the average quality or performance by removing inferior individuals. In connection with seeds, the increase of viability and vigour of a seed lot by removal of small, immature, empty and otherwise inferior seeds from a seed lot. In connection with seed sources, the culling or roguing of inferior phenotypes to improve the genetic quality.

**Ultra-dry storage:** Seed storage in conditions where low temperature storages are not available. Here seed is dried to <5% moisture and equilibrium at 10-12 RH% and temperature at 20°C

## V

**Vigour:** The vitality or strength of germination especially under unfavourable conditions.

## W

**Wet seeds**: This type of seed like tomato, peppers, egg plant, capsicum, watermelon, cucumber, bitter gourd, bottle gourd or pumpkin, where the seed is attached to flesh of the fruit and seed separation will be based on fermentation principle

**Winnowing:** An ancient technique used to clean seeds moving air from a fan or breeze is used to separate heavier seeds from lighter chaff.

## Study Questions

1. List the organizations dealing with seed research all over the world

| S.N. | Organization | Location and Country | Specialization |
|---|---|---|---|
| e.g | Directorate of Seed Research DSR | Mau, UP,India | Seed Research and training |

2. List the websites giving information on seed related aspects-specify what information they give or deal with.

| S. N. | Website | Location and Country | Specialization |
|---|---|---|---|
| e.g | www.seedquest.com | USA | Global Information Services for Seed Professionals |

# Practical – 1

# Visit to a Shop Floor of a Seed Processing Unit

**Visit to a seed processing unit and attending the following questionnaire to know firsthand information on operations of shop floor**

**Study Questions**

1. List the equipments of seed processing and their working principles.
2. List the equipments of seed storage and their working principles.
3. List the equipments of seed testing and their working principles.

## Fill the following table

| Sl. No. | Equipment | Principle of working | Purpose |
|---|---|---|---|
| | | | |
| | | | |
| | | | |
| | | | |
| | | | |
| | | | |
| | | | |
| | | | |
| | | | |
| | | | |
| | | | |

# Practical – 2

# Seed Extraction Methods
# Activity: Seed Extraction of Tomato

## 1.1 Problem

1. Compare the seed quality at different physiological maturity stages
2. Compare the seed recovery in acid and fermentation method

## 1.2 Materials

- Tomato fruits – 4 batches
- Plastic bucket/ Beaker
- Water

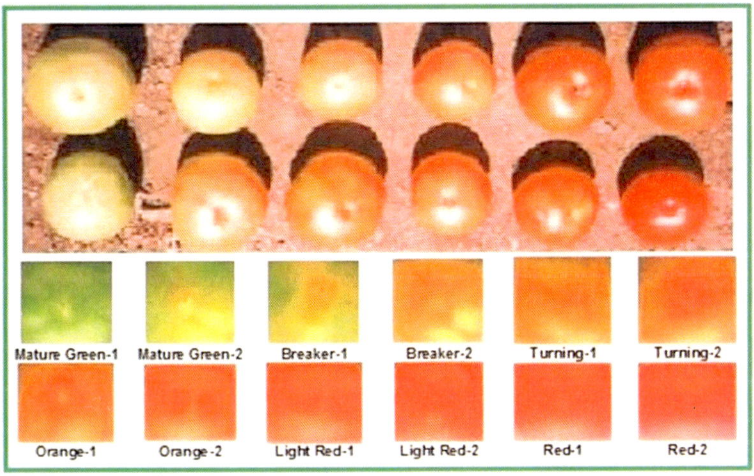


Fig. 1: Seed physiological maturity chart

## 1.3 Procedure

1. Seed quality at different physiological maturity stages
2. Seed recovery in acid and fermentation methods

### *Acid method*

1. Squeeze 1 kg of fruits.
2. Add hydrochloric acid @ 10 ml/kg of pulp.
3. Frequently stir it for 30 minutes.
4. After 30 minutes remove the floaters and wash the settled seeds with water for 3 to 4 times.
5. Decant the water and shade dry the seeds. Later sundry the seeds for 2-3 h.
6. Weigh the dry seed and calculate the seed recovery percent.

### *Fermentation method*

1. Squeeze 1 kg of tomato fruits and add enough quantity of water.
2. Allow the fermentation of the pulp without disturbing for 24-48 hours.
3. Remove the pulp that floats after decay.
4. Wash the settled seed repeatedly and decant the water.
5. Shade dry the seeds initially, later sundry for 2-3 h.
6. Weigh the dry seed and calculate the seed recovery percentage.

**Fig. 2:** Tomato seed extraction by fermentation method

## 1.4 Observations

### *1.4.1 Physiological maturity*

| Stage of phy. maturity | Germination% | SVI |
|---|---|---|
| | | |

## Key to tests

| S. No | Test | Test Developed By | Significance and Formula |
|---|---|---|---|
| 1. | Volume based Imbibitions Time (Hr) | Leopold (1983) | Time taken for 2mm emergence of primary root from the seed coat.Kt=-log= $(a_{max} - a)\ a_{max}$ Where $a$ is the volume at tune $t$, is the volume at full hydration(24 h) and $k$ is swelling co-efficient |
| 2. | Germination (%) | Nichols *et al.* (2009) | Seeds were considered to have germinated when radicles reached at 5 mm length Germination percentage = (No. of germinated seeds Total No. of seeds sown) x 100 |
| 3. | Rate of Germinatior | Agarwal (2004) | Rate of Germination = Σ (*ni/Di*) *ni* is the number of germinated seed on day *i* and *Di* is the number of days after beginning the experiment. |
| 4. | Mean Germination Time | Ellis and Roberts (1981) | MGT = Σ (Dn)/ ΣnWhere *n* is the number of germinated seeds or emerged seedlings on day D and D is the total number of days counted from the beg tuning of germination. |
| 5. | Mean Daily Germination | Czabator (1962) | MDG = Total number of germinated seeds Total number of days |
| 6. | Peak Value | Czabator (1962) | PV = Highest seed germinated Number of days |
| 7. | Germination Value | Czabator (1962) | GV= Peak value X Mean Daily Germination |
| 3. | Emergence Rate Index | Adetimirin (2003) | ERI estimates the total number of days for 100% emergence to occur |
| 9. | Shoot Length (cm) | Agarwal (2004) | Ten normal seedlings of each replication were taken and shoot length was measured in centimeter computed |
| 10. | Root Length (cm) | Agrawal (2004) | Ten normal seedlings from each of die replication from germination test were removed carefully and used for measuring root length. Root length was measured in centimeters from die collar region to die tip of main root with die help of scale. The average of ten seedling roots was computed and expressed in centimeters. |

| | | | |
|---|---|---|---|
| 11. | Seedling Vigour Index | Abdul Baki and Anderson (1973) | The seedling vigor index (SVI) = (Seedling length (cm) X Germinated percentage)/100 |
| 12. | Seedling Dry Weight (mg) | Abdul and Anderson(1973) | Dry weight of seedlings were recorded after drying diem in die hot air oven at 700C for 4S hours and they are cooled by keeping in a desiccators hour and die average dry- weight of die seedlings was expressed in milligrams. |
| 14. | Relative Dry Weight | Czabator (1962) | Increase of dry weight of seedlings over die control |
| 15. | Seed Viability (%) | Perry (1987) | Seed viability was determined by a test using 400 seeds, placed in three separatereplications of 100 seeds |
| 16. | Usable Transplants (%) | Hartley *et al.* (2001) | Percentage of healthy, well grown, saleable seedlings or transplants obtained per gram of seed at 10-14 days after sowing. |

### 1.4.2 Method of extraction

| Method | Seed Recovery % |
|---|---|
| Acid method | |
| Fermentation method | |

## Study Questions

1. List the seed extraction/threshing methods in different crops and seed moisture levels at harvesting (10 crops each in cereals, pulses, oilseeds and vegetables).

| S. No | Crop | Seed extraction method | Seed moisture at harvesting |
|---|---|---|---|
| **A Cereals** | | | |
| 1. | | | |
| 2. | | | |
| 3. | | | |
| 4. | | | |
| 5. | | | |
| 6. | | | |
| 7. | | | |
| 8. | | | |
| 9. | | | |
| 10. | | | |
| **B Pulses** | | | |
| 1. | | | |
| 2. | | | |
| 3. | | | |
| 4. | | | |
| 5. | | | |
| 6. | | | |
| 7. | | | |
| 8. | | | |
| 9. | | | |
| 10. | | | |

**C Oil seeds**

1.
2.
3.
4.
5.
6.
7.
8.
9.
10.

**D Vegetables**

1.
2.
3.
4.
5.
6.
7.
8.
9.
10.

# Practical – 3

# Seed Quality Up-gradation

Seed processing is specialized activity and also a tool to improve seed quality further to an acceptable standard. Seed quality parameters like germination, vigour or genetic purity can be improved using processing. Say a seed lot having a minimum standard of 90% germination, but the competition in the market offering seed quality 92 or 95 % then to meet the market demand we can process the seed lots using re-milling or re-conditioning techniques. Also sometimes simple techniques like liquid density separation which uses specific density of the seed to separate good seed from inferior quality can be considered.

## Method-1: Processing using specific gravity separator

### No of lots-4 (Seeds)

1. Bulk lot-after harvest (M1)
2. Hand cleaned lot (M2)
3. Air screened lot (M3)
4. Reject (M4)

**Procedure**: The lots should be upgraded by using specific gravity separator. Then each to test for germination and seedling vigor index before and after the up gradation.

### *Before up gradation*

**Table 1:** Seed quality before up gradation

| | T* Germination% | | | | SVI[$] | | | |
|---|---|---|---|---|---|---|---|---|
| | R-1 | R-2 | R-3 | M[#] | R-1 | R-2 | R-3 | M |
| M1 | | | | | | | | |
| M2 | | | | | | | | |
| M3 | | | | | | | | |
| M4 | | | | | | | | |

## *After up gradation*

**Table-2:** Effect of up gradation on seed quality

| T* | Germination% | | | | SVI[$] | | | |
|---|---|---|---|---|---|---|---|---|
| | R-1 | R-2 | R-3 | M[#] | R-1 | R-2 | R-3 | M |
| M1 | | | | | | | | |
| M2 | | | | | | | | |
| M3 | | | | | | | | |
| M4 | | | | | | | | |

* T-Treatments #M-Mean

$ SVI-Seedling Vigor Index

## Method-2: Liquid density separation by using salt solutions

**No of lots-4 (Seeds):** Bulk lot-after harvest (L1 to L4))

### Procedure:

1. Prepare salt solutions of different conc.
2. Seeds to be immersed in salt solutions
3. Test the seed quality using routine tests

**Fig.. 1** LDS using salt solutions

## Treatments

**Table 3:** Effect of LDS with salt solutions

| T* | Germination% | SVI$ |
|---|---|---|
| 5% | | |
| 10% | | |
| 15% | | |
| 20% | | |
| DW & | | |
| Control | | |

* T-Treatments-different salt solutions & Distilled water

## What other options?

- Try different organic and in-organic solutions
- Use common household items like vinegar, edible oils to increase the density
- List out some commonly available liquids and calculate the specific densities by using formula:

$$\rho \text{ (Rho)} = \frac{\text{m (mass)}}{\text{V (volume)}}$$

## How?

- Find the Mass of an empty Measuring Cylinder.
- Add a certain Volume of Liquid.
- Find the Mass of the Measuring Cylinder and Liquid. Calculate the Mass of Liquid.
- Mass of Liquid = Mass of Measuring Cylinder and Liquid – Mass of empty Measuring Cylinder -Calculate Density of Liquid (Unit- g/cm3 or $kg/m^3$

## Study Questions

1. List out specific densities of the organic and inorganic media which have potential for seed separation (10 each)

| Sl No | Media | Specific density-g/cm$^3$ | Type-organic or inorganic |
|---|---|---|---|
| **A Organic compounds** | | | |
| e.g | Pure water | 1.00 | Organic |
| 1. | | | |
| 2. | | | |
| 3. | | | |
| 4. | | | |
| 5. | | | |
| 6. | | | |
| 7. | | | |
| 8. | | | |
| 9. | | | |
| 10. | | | |
| **A Inorganic compounds** | | | |
| e.g | Di-ethyl ether | 0.71 | Inorganic |
| 1. | | | |
| 2. | | | |
| 3. | | | |
| 4. | | | |
| 5. | | | |
| 6. | | | |
| 7. | | | |
| 8. | | | |
| 9. | | | |

## Practical – 4

# Estimation of Physical Properties of Seeds

The physical properties of seeds are a basic activity for designing processing equipments and precision planters. Since these characters vary with crop or variety, one needs to have the data handy to be used on shop floor. This data also helps to quantify processing efficiency, presence of dormancy and suitability for seed treatments for all existing and new crops or varieties.

**Materials**: Gum Guar Seeds, Leica EZ for Image analysis, Screens of different sizes

### Physical properties to be quantified:

1. General photograph of the sample in Leica EZ
2. Length (cm)
3. Width (cm)
4. Thickness (cm)
5. Diameter (cm)
6. 1000 or 100 Seed Weight (g)
7. Seed moisture (%)
8. Hilum Length-cm
9. Arithmetic diameter-cm
10. Geometric diameter-cm
11. Volume-$cm^3$
12. Sphericity (*SÕ*)
13. Bulk density, *ñ*b(kg/m3)

14. True density, *ñ*s (k g/m3)
15. Porosity, *å*(%)
16. Angle of repose, q(°)
17. Surface area, *S*a(mm2)
18. Specific surface area, *S*s (cm2/cm3)
19. Hydration capacity, *H*c(g/seed or kernel)
20. Hydration index, Hi
21. Swelling capacity, *S*c(ml/seed or kernel)
22. Swelling index, Si

*Note: Refer Worksheet -1 for formulae to calculate*

**Fig. 3:** Gum guar seeds and plant

**Crop:**

**Variety:**

| Physical Properties | R-1 | R-2 | R-3 |
|---|---|---|---|
| Length-cm | | | |
| Width-cm | | | |
| Thickness-cm | | | |
| Diameter-cm | | | |
| 100 Seed Weight-g | | | |
| 1000 Seed Weight-g | | | |
| Seed moisture-% | | | |

Hilum Length-cm
Arithmetic Diameter-cm
Geometric Diameter-cm
Volume- $cm^3$
sphericity $(S\Phi)$
Bulk Density-$\rho$b(kg/m3)
True Density-$\rho$b(kg/m3)
Porosity $\varepsilon$(%)
Angle of repose- q(°)
Surface area-*S*a(mm2)
Specific Area-*S*s (cm2/cm3)
Hydration capacity- *H*c(g/seed or kernel
Hydration index, Hi
Swelling capacity- *S*c(ml/seed or kernel)
Swelling index-Si

# Practical – 5

# Seed Blending

## Methods in Seed Blending

For good seed blending there are three methods and details of which given as below:

1. **Mechanical Seed Blending using a blender**: In this method seed blending is done using a mechanical blender or mixer. Here to achieve uniformity all the constituent seed lots will be mixed in a blender. The mixing involves movement of particle s induced mechanically or by gravity. The two seed lots are fed in to the feeder bins simultaneously and circulated which ensure uniform mixing of these lots. The constituent seed lots held in large storage bins are measured electronically and introduced in to the blender at required proportions. The blended seed then automatically packed as bulk which will be used for packing smaller pockets or tins later. A typical mechanical blender operation is given in Fig.6.1

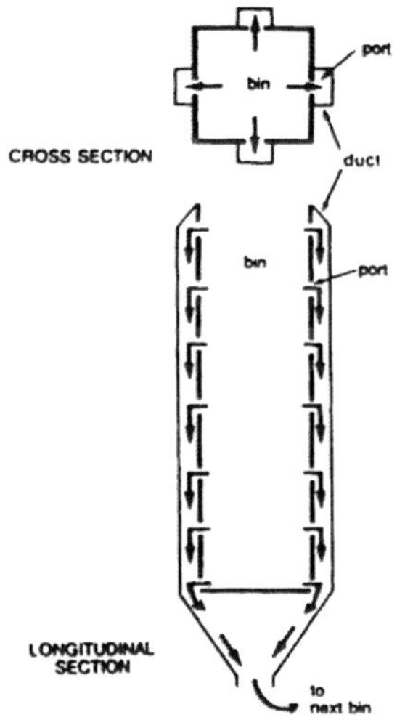


**Fig. 1:** Operation of a mechanical seed blender

2. **Manual blending based on PLS (Pure Live Seed)**: The seed blending in the absence of large mechanical blender can be done following certain principles (detailed considerations given at the end of this chapter). One of the approach is based on pure live seed or PLS. This type of blending is useful when two lots have a similar quality and mostly similar quantity (+ or – <5%). Here is an example for this method:

**Step-1:** Calculation of PLS-Pure live seed can be calculated using the following formula:

$$PLS = \frac{Purity \times Germination}{100}$$

**Step-2:** Blending of lots : If the minimum PLS is 85%

**Lot A:** Seed lot having 98 % purity and 90 % germination gives 88.2 PLS

**Lot B:** Seed lot having 98% purity and 88 %germination gives 86.2 PLS

Then the blending of these two lots gives acceptable PLS

**Seed Blend AB**= (88.2+86.2)/2=87.2 PLS (accepted for blending)

3. **Manual blending based on germination %**: In this case the germination is considered when the seed lots to blend have different quality and quantity. In principle only two lots of having germination difference of more than 10% are accepted for blending. This approach is widely followed and useful for improving the quality of smaller lots.

**Step-1:** let us take minimum standard required is 90% of germination

**Lot A:** Seed lot having 50 kg weight and 95% germination

**Lot B:** Seed lot having 25 kg weight and 80% germination

**Step-2**: Blending of the lots

**Lot A:** 50kg X 95% of germination=4750.......1

**Lot B:** 25kg X 80% germination=2000...........2

Then add 1+2=we get 6750 and divided by total quantity i.e 75 kg we get 90 which is acceptable for blending since it is the standard fixed for the blend. We have improved seed lot quality of lot B from 80% to 90% by blending with bigger lot A having better germination quality.

## Study Questions

1. Workout whether blending is permissible in the following Tomato lots-use all the methods

| Germ% | Purity % | | Qty-Kgs | | Std | |
|---|---|---|---|---|---|---|
| | Lot-1 | Lot-2 | Lot-1 | Lot-2 | Lot-1 | Lot-2 |
| 85 | 94 | 88 | 90 | 25 | 44 | 92 |
| 88 | 96 | 85 | 95 | 45 | 28 | 94 |
| 89 | 94 | 84 | 89 | 28 | 33 | 90 |
| 86 | 94 | 92 | 87 | 15 | 65 | 95 |

# Practical – 6

# Determination of Mechanical Injury After Processing

Mechanical damage is inflicted on seeds during harvesting, threshing drying, processing and handling. Injury may occur from any kind of physical abuse. Threshing abuse probably is most serious but, seed conditioning and handling also contribute to seed injury. High temperatures during drying or drying too quick or excessively can cause injury. Susceptibility to mechanical injury increases as the moisture content of the seed decreases. Effect of all individual impacts received by the seed appears to be cumulative in total seed damage. Cereals are largely immune to mechanical injury presumably because of the protective outer structures, the palea and lemma. Large sized legumes are more suscepitable. Small seeds tend to escape injury during harvest and seeds that are spherical tend to suffer less damage than elongated or irregularly shaped ones.

## Symptoms

1. Splits or cracks in seed coat which is easily visible.
2. Internal damage, visible only after seed germination. Injuries close to vital parts of the embryonic axis or near the point of attachment of cotyledons to the axis usually encounter the most rapid loss of viability.
3. Microscopic breaks especially on seed coat, reduces performance of seeds and increases susceptibility to microorganisms.
4. Cryptic (hidden or internal) injury is more physiological and reduces the seed germination and vigour. Abnormal seedlings resulting from mechanical damages are i. Under developed or brown cotyledons. ii. Twisted epicotly or hypocotyls region; root tip is blunt. iii. Improper development of essential structures of the seedling.

## Methods to Asses mechanical injury

### 1. Ferric Chloride Test

The seeds placed in 20% solution for 15 min, if mechanically injured become stained black while, live tissues become stained pink. Because of the rapidity of this test, this can be used as a check test while adjusting the processing machineries.

### 1. Tetrazolium Test

The seeds are placed in 2 to 2.5% 2,3,5 triphenyl tetrazolium chloride solution for 1-4 h in dark after which, the seeds are observed. The deep red staining indicates mechanical injury.

### 2. Malachite Green Test

The seeds are soaked in 5 to 10 % solution of malachite green for 20 min. The damaged portion becomes dark green, while other parts are yellow in colour.

### 3. Indoxyl Acetate Test

It helps in identification of mechanical injury in large seeded legumes. Here the seeds are soaked for 10 sec in 0.1% solution prepared in 95% ammonia and later air dried. The lesion of the seed coat becomes purplish green.

## Activity: Assessing Mechanical Injury

### Materials

- Malachite green 10% solution.
- Soybean seeds.
- 250 ml beaker.

### Procedure

1. Take 10% solution of malachite green in a 250 ml beaker.
2. Soak the seeds in the solution for 20 min.
3. Remove the seeds from the solution and rinse in water.
4. Observe the seed coat for presence of green lines, which indicates mechanically, injured regions.

# Practical – 7

# Basic Seed Processing Equipments

The processing operations are carried out based on the principle of physical differences found in a seed lot.

**Table 1:** Physical properties of seeds and machines based on them

| Physical difference | Suitable machineries |
|---|---|
| Seed size – varied from small to bold | Air screen cleaner cum grader |
| Density- ill filled, immature to well matured light weight to dense seed | Specific gravity separator |
| Shape – round to oval and different shapes | Spiral separator |
| Surface texture – smooth to wrinkled and rough | Roll mill / dodder mill |
| Colour of the seed – light colour to dark colour | Electronic colour sorter |
| Conductivity of seed – low to high | Electronic separator |

## A. Study of Air Screen cleaner

**Principles :** Basic and cleaning is done by air screen cleaner which separates undesirable materials and seed from desirable seeds on he basis of difference in seed size, seed shape and seed weight using screen and air blast in 3 stages.

Such as

a) **Aspiration :** Removal of light and chaffy seed materials

b) **Scalping :** separation of the seeds, which are dropped through screen opening but, the larger material (rash, colds, etc.,) retained on top of screen are carried over into a separate spout

c) **Grading :** Separation of the good seeds which ride over the screen opening into a separate pile, while smaller particles pile, while smaller particles (under sized, cut, shrived, broken seeds, sand/soil particles) passed through the screen-mesh.

**Working Procedure:** In four screens cleaner the screens do the following job in sequence. First screen-Scalping., Second screen-Grading, Third screen-close scalping an dfourth screen-closed grading. Seeds are uplifted through Bucket elevator to feed hopper where they pass through the upper air, which removes light chaff and dust by air blast. The seed to be processed is passing on top screen, where rough scalping done by removing the larger materials riding over the screen. The seeds pass through screen will be subjected to the rough grading, the good seeds ride over the second screen were lifted on third screen for rough scalping. Particles larger than the seed will be collected from separate spout. The good seed fall from this screen will graded finally where smaller seeds including chaff, shriveled, broken seeds were removed from other spout.

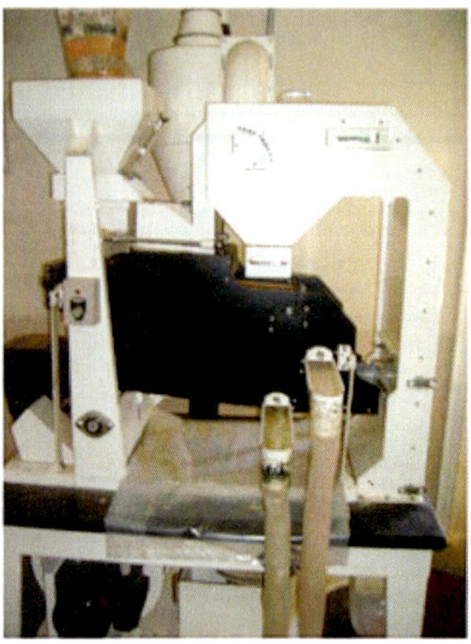

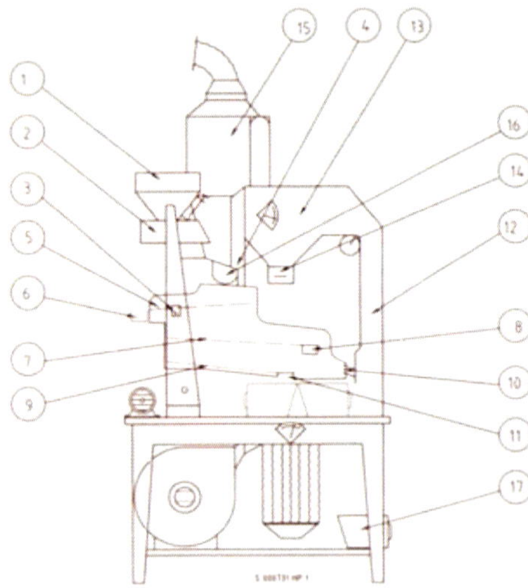

Here the light and inferior quality seeds are get separated from good seed. The product to be cleaned poured into the inlet hopper (1) where the vibratory feeder (2) leads it to the sieveboat. The Triac control (3) sets the capacity. After leaving the vibratory feeder, the product passes the pre-aspiration (4) where light particles are removed. The product is led to the short scalping screen (5) where large particles are removed and led out through chute (6). The overflow continues to the first grading screen (7) where the overflow is led out through chute (8). The through flow continues to the second grading screen (9). The through flow (sand and small particles) from the second grading screen is led out through chute (11). The overflow is led out through outlet (10) and up into the final aspiration channel (12) where the last light particles are removed. These particles settle in the drawer (14) under the aspiration chamber (13). Dust is sucked with the air to the cyclone (15) where the dust particles settle in the glass (16). The prime product leaves in the bottom of the final aspiration channel and settles in the drawer (17).

Screens are selected according to shape of the crop seed to be cleaned. For round sees- Top screen hole should be round and bottom screen hole slotted/round. While for oblong seeds both top and bottom screen holes should be oblong. For lens shaped seeds top screen hole is oblong/rectangular and bottom screen hole is round.

1. **Air blast/Fan :** Most air screen machines provided with two air systems. The upper one remove light chaff and dust from seed before they reach the screen and is regulated by an air adjustable dampers in the air passage and the lower air blast removes light seed and trash not removed by the air passage and the lower air blast removes light seed and trash not removed by the upper air on screen. For efficienct cleaning, the lower air blast should be strong enough to blow out of materials/chaff of good sed size.
2. **Clay crushing Rolls:** Two large diameter rubber head rolls are installed to crush clay lumps without damaging seeds. The rules are adjustable to various opening widths and may be left apart and inoperative when not neede.
3. **Brushes:** Brushes travel mechanically back and forth under each screen to sweep the undersized clean and keep the perforations open. The cleaning efficiency of screen is directly related to the percentage of perforations that remain open during operation.
4. **Shoes:** The vibrating or shaking sections of the machine to which the screens are fitted are called shoes. A shoe usually has fittings for two screens, one for scalping and other from grading. A machine may have on eor two shoes depending upon its size. The shoes can be inclined to allow the seeds to flow over the screen.

## Other Adjustments to increase efficiency

a. Slope of the screen

b. Volume of the air blast

c. Rate of vibration of screens

d. Types of screen

## B. Study of Gravity seperator

The gravity separator, the stoner, the aspirators and the pneumatic separator are all designed to make separations by differences in weight or specific gravity of seed.

**Principle:** threshed seed processed with basic cleaner will be upgraded by subjecting to specific gravity separator in which good seeds were separators from light seeds or chaff/shriveled seeds based on density by adjusting air flow stream under deck of the machine.

**Materials:** Seed material, specific gravity separator and bags

**Working Procedure:** As seed flow on the gravity separator they enter a column of were coming up through the porous surface of the deck. The pressure or terminate velocity of the air rising through the deck can be controlled very closely. Once the seeds have been stratified, these vertical zones will be over to different parts o fthe deck so that that can be discharged through different spouts. Specific gravity separator consists of perforated deck with an air stream moving through it. Seeds are fed on the deck and into the air stream. The air stream is adjusted to lift light seed while heavier seed lies on the back surface. Then stratified layer of seeds is separated so that seed to one discharge spout and heavier seed to the deck. As the deck reciprocates it tosses heavy seed in an uphill direction. This motion continues until heavy seeds are carried to the end of the deck and got out to a discharge spout.

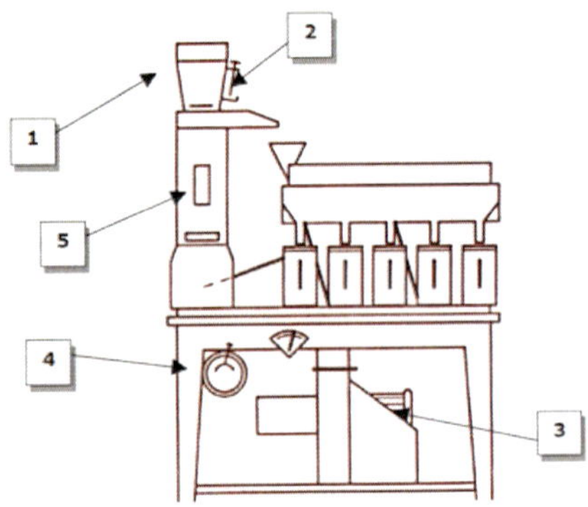


Here the separation is in phased manner from ill filled, immature to well matured light weight to dense seed. The product to be graded enters through the hopper (1) fitted with adjustable feed gate (2). The motors (3) and (4) are started. The vibrator (5) is started and the material now has to be fed in a steady flow so that it covers the entire table.

**Adjustments:** The gravity separator is a versatile machine an dcan accomplish a wide range of separations because it has many adjustments, which allow the separator to control the separating action precisely. Since each adjustment affects the action of the others, all adjustments must be coordinated or blended together to produce a sharp separation.

***There are five main adjustment affects the performance of the machine***

1. **Feed rate:** A constant and uniform rate of feed is essential to maintain the uniform bed of seeds on the deck. The thickness of the should be sufficient to allow the most effective stratification and separation. Two much feed rate will not ensure uniform separation.

2. **Air control :** It is the basic adjustments so that the seed mass should fluidizes and flows freely and the heavy seeds lies on the deck and lighter seeds are lifted into upper layer of the seed mass. *Excessive air for heavily seeds made them to flow into the layer of the light seed and destroy stratification. This is characterized by bubbling in the seedbed and discharge of heavy seeds with the light seed. *Insufficient air fails to lift the light seeds aboe the deck surface and they fails to stratify the seeds. This causes the seed to 'dead' on the deck and light seeds will discharge with heavy seeds.

3. **End slope:** End slope from the seed hopper to the discharge end, controls to speed at which seed move deck and thus length of time obtained to remain on the deck. When the differences between seeds to be separated are slight. The deck should have relatively flat end slope to hold seeds on deck longer period. Crop seed and contaminants that differ greatly in specific gravity will stratify and separate quickly. So end slope can be increased to move the seed on the deck rapidly and increase capacity.

4. **Slide slope:** It is tilt of inclination of the deck from low side to the high side of the discharge end. Slide slope creates an inclined surface over which stratified seedbed must flow to reach the discharge end. This allows the light seed layers riding on the cushion of air to slide down to the low side of the deck, while deck oscillation moves heavy seed uphill to the opposite side of deck.

5. **Deck oscillation speed:** The motion of the deck causes heavy seed to move towards the high side as they flow from the feed area to the discharge end. Decrease in deck speed causes heavy seeds to discharge lower on the deck.

## Machinery adjustments

**Air adjustment** - for stratification seed

**Deck** - to causes the mass of seed to cover the deck completely and discharge evenly along the lowest of the deck.

**Speed of deck** - increase in speed – more seed mass towards high end decrease in speed- more seed mass towards the low end.

## Study Questions

1. List the companies operating in India and the rest of the world in seed processing equipments

   e.g: Indosaw (India) and
   Seed Processing(Holland)

| Sl no | Name of the Company | Country | Any 5 of their | Website address |
|---|---|---|---|---|
| e.g | Indosaw | India | Seed grader | www.indosawagri.com<br>equipments<br>Seed -analyser |

# Practical – 8

# Advanced Equipments of Processing

## 9.1 Magnetic seed separator

These machines use a magnetic force to separate damaged seeds or other particles from the good seeds. This is used to solve special problems in seed lots that can not be taken out with our other basic seed cleaning equipment. Most of the time it is used for vegetable and field seeds that have problems in seed damage (fish mouth in cucumber or melon seeds / cracked radish or cabbage seeds), insect damage (beetle holes in cotton seeds) or contamination with soil or plant parts (white plant parts in onion seed) etc.

**Fig .1:** Magnetic seed separator for problematic seeds

## 9.2 Seed mixing/blending machine

These machines are used for mixing/ blending of different seed lots to make it a homogeneous lot.

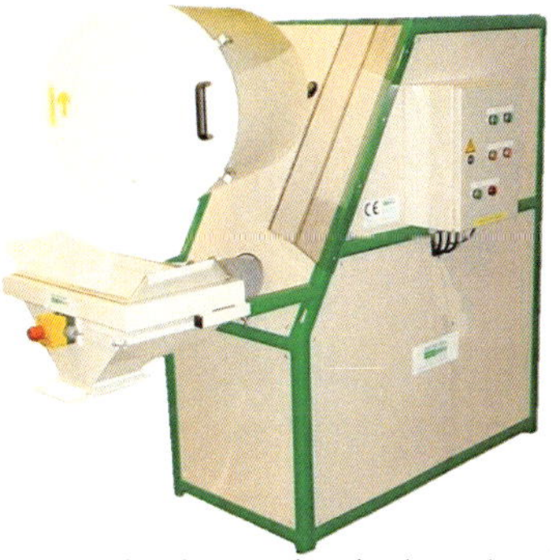

**Fig. 2:** Mechanical seed blending machine

## 9.3 Blower

The use of this as testing machine used for the classification of a sample according to the aerodynamic behaviour (floating speed) of the particles in an air stream The sample is placed in a sample tray with fine mesh in the bottom. Air is blown through the mesh and continues up through the vertical separation pipe. The sample is then divided into two fractions – one in the settling chamber and one in the sample tray (pure seed collection).

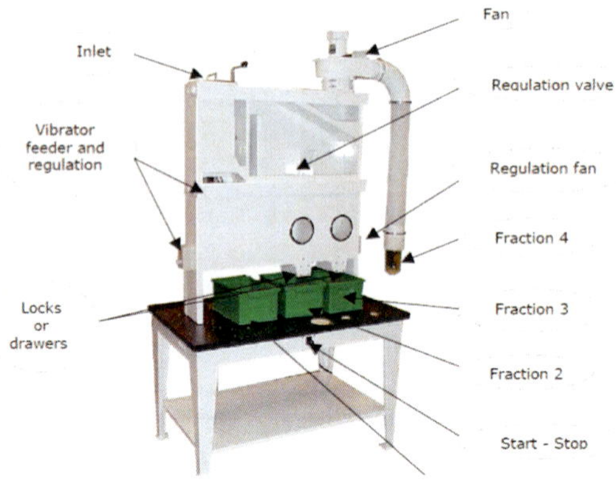


**Fig 3.** Aspirator or seed blower to separate seeds on their aero-dynamic properties

## 9.4 Feed Hoppers

For all machines the feed hoppers play an important role in easy flow of material and also reduce the labor requirement. Also for bigger machines along with belt conveyors hoppers will work more efficiently. These can be of different capacities based on the type of machine in use and volume of the material.

**Fig. 4:** Feed hoppers for holding and movement of seed in to the processing machines

## 9.5 Screens

For all basic and advanced seed cleaning machines screens are important parts which need to be used based on the type of seed. There will be round (R) or slotted (S) screens and of different sizes measured in mm for crops.

**Fig. 5:** Screen types to separate seeds based on shape

## 9.6 Seed Winnower

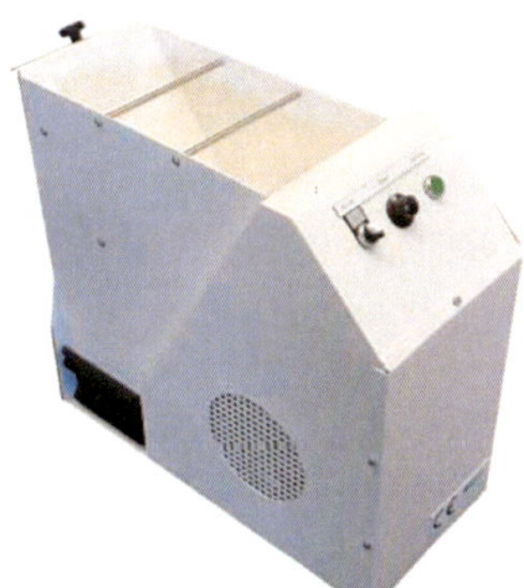

**Fig. 6:** Seed winnower to clean smaller seed lot

The winnower is used to utilise the differences in specific weight between the seeds and other parts. A rough air separation separates the seeds from the husks. It is very suitable for cleaning small seed lots.

## 9.7 Seed disinfecting or treatment equipment

To eliminate bacterial contamination or seed borne diseases. It can be used as an imbibition unit for the uptake of moisture and for hydration before priming or as a hot water treatment unit. The system uses seed drums to carry the seeds. The seeds are treated in a very uniform and reproducible way.

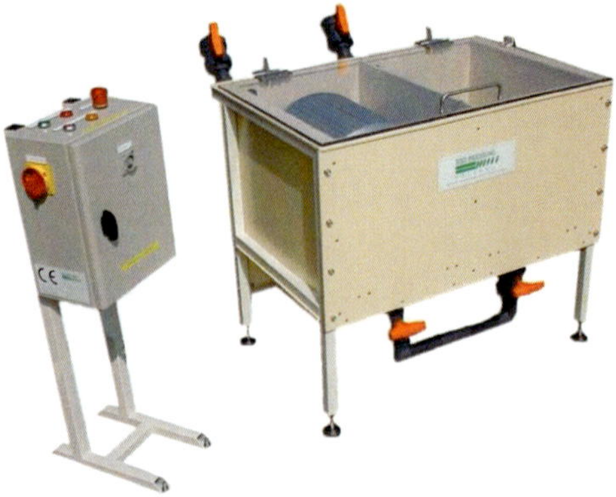

**Fig. 7:** Mechanical seed sanitation and dis-infecting equipment

## 9.8 Seed osmotic priming unit

This systems use low water potential osmotica, such as polyethylene glycol (PEG) or slats and air (oxygen) to enhance germination of a seed lot. The goal is to reach a faster and more uniform germination.

**Fig. 8:** Osmotic priming unit for seed enhancement

## 9.9 Seed density separation unit

This machine is used for accurate seed density grading using liquids (water or liquids with certain osmotic pressures). During separation the seeds should keep the same moisture level content.

**Fig. 9:** Liquid density separation unit

## 9.10 Seed air separator

This air separator is used after the seeds are threshed. By means of air, dust and small plant parts are separated from the seeds. By using this machine a lot of waste material is removed before the fine cleaning process takes place.

**Fig. 10:** Mechanical seed air separator

## 9.11 Pre-cleaner

This machine is used after the seeds are threshed. This machine separates dust and capsule parts in a fast and easy way, before the fine cleaning process.

**Fig. 11:** Pre-cleaner to separate chaffy and denser inert matter off seed

## 9.12 Seed head elevator and dosage belt

This elevator is used for transportation and dosing of seed heads or umbels into the inlet of our seed thresher.

**Fig. 12:** Seed overhead elevator and dosage belt

## 9.13 Seed Extractor

With this seed drill you can manually extract seeds from cucurbita fruit crops.

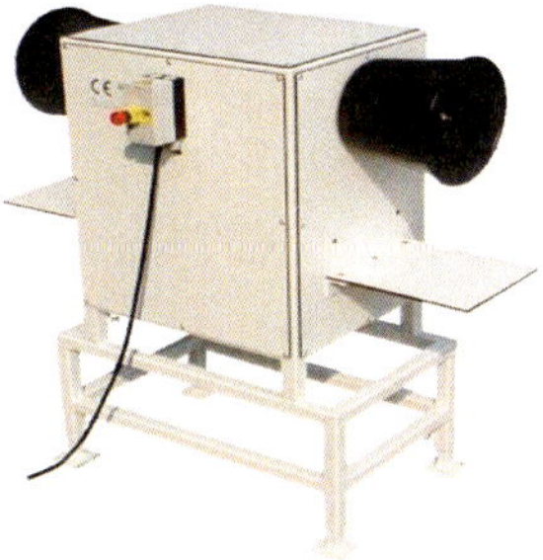

**Fig. 13:** Seed extracting equipment from fleshy fruits (tomato or cucumber)

## 9.14 Seed rinsing channel

This rinsing channel is used to separate heavy seeds from lighter seeds and fruit parts/skins after a seed extraction process.

**Fig. 14:** Seed rinsing chamber to remove fruit and seed inert matter

## 9.15 Seed washing and sieving unit

This unit is used to separate small quantities of seeds from larger quantities of pulp and skin parts after the seed extraction process.

## 9.16 Seed centrifuge

This unit is used in combination with our large seed drum. After seed extraction or any kind of wet treatment the centrifuge is used to get rid of moisture that is around the seeds, which makes the drying process that follows shorter.

**Fig. 16:** Centrifuge to remove excess moisture off the seed material

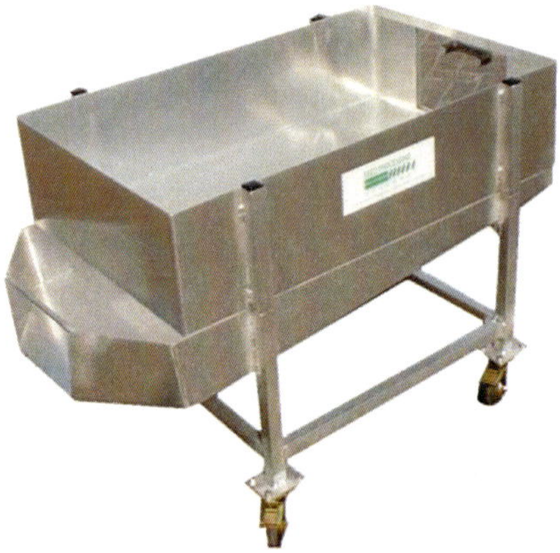

**Fig.15:** Seed washing and sieving unit

## Study Questions

List out the seed processing equipments available in India with the principle of operation

# Practical – 9

# Standardization of Screen Sizes for Seed Processing

Seed size is an important parameter of seed vigor as it in influences the performance of seed in soil. Seed bulk at harvest contains a wide range of seed sizes but these may not all be of equal value for sowing. Farmers have always realized the necessary of using uniform seeds for good viability to obtain high emergence and growth. This will also help in getting optimum plant population and higher yields.

The influence of seed size on seedling vigor and crop productivity has already been proven since the days Hannaford who was considered as father of seed processing. To achieve uniformity in seed size this optimum sieve/screen size is very critical for optimum seed recovery and economical seed yield. Though the screen sizes are standardized for different crops but vary with different varieties. The shop floor should continuously explore and fine tune screen sizes for different varieties.

In this practical we will standardize the screen sizes for new pigeon pea cv. TS3R. The methods employed here can be used for different crops. The seed quality parameters like seed recovery %, graded seed yield (q/ha), germination %, 100 seed weight-g and field emergence will be measured.

## General Rules in Standardization

- There will be two/three or four screen air screen cleaners and graders based on the requirement.
- Three screen graders are commonly used
- The three screens are TOP-MIDDLE-BOTTOM.

| Screen | Purpose |
|---|---|
| TOP | Scalping |
| MIDDLE | Remove broken seed |
| BOTTOM | Retain good seed |

To make this experiment simpler, we design an experiment only for the bottom screen keeping all the other screens as standard

| | | |
|---|---|---|
| Crop | : | Pigeon pea cv. TS3R |
| Top screen | : | 9.00 mm-Round |
| Middle screen | : | 3.25 mm-Slotted |
| Bottom screen | : | As below |

| Screen size-mm | Seed recovery % | Graded seed yield-q/ha | 100 seed weight-g | Germination % | SVI |
|---|---|---|---|---|---|
| Un-processed seed | | | | | |
| 3.25 | | | | | |
| 3.50 | | | | | |
| 3.75 | | | | | |
| 4.00 | | | | | |
| 4.25 | | | | | |
| 4.50 | | | | | |
| 4.75 | | | | | |

*Criteria for seed quality parameters*:

**Seed recovery %** =

$$\frac{\text{Wt. of graded seeds}}{(\text{Wt. of graded seeds} + \text{Wt. of rejected seeds})} \text{ X } 100$$

**Grades seed yield (kg/ha)** =

Processing recovery % X Unprocessed seed yield /ha

## Study Questions

List out top and bottom screens prescribed for seed processing by seed certification agencies

# Practical – 10

# Effect of Screen Sizes on Seed Quality

This practical in continuation of the practical no.10 explore the effects of screen sizes on different varieties. The screen sizes vary not with crops but with varieties also. To know the effect of screen sizes among the varieties we need to take the top and middle screens as standard or as recommended. These methods can be applied to any crop and to different varieties.

### *Procedure*

1. Take at least 100 kgs of seed of each variety
2. Keep the top and middle screens as standard
3. Change the bottom screens
4. Collect seed samples from the outlets of the machines after each run with specified bottom screen
5. Test the quality parameters for different varieties
6. Good test results found in a particular screen sizes can be considered suitable for that variety
7. Keep one unprocessed seed sample as control

**Effect of screen sizes on different quality parameters of sunflower**

**Variety: 1**

| Screen Size-mm | Seed recovery % | 1000 seed wt-g | Physical purity-% | Germ-% | SVI |
|---|---|---|---|---|---|
| Un-P | | | | | |
| 1.85 | | | | | |
| 2.50 | | | | | |
| 2.80 | | | | | |
| 4.00 | | | | | |

**Variety: 2**

| Screen Size-mm | Seed recovery % | 1000 seed wt-g | Physical purity-% | Germ-% | SVI |
|---|---|---|---|---|---|
| Un-P | | | | | |
| 1.85 | | | | | |
| 2.50 | | | | | |
| 2.80 | | | | | |
| 4.00 | | | | | |

**Variety: 3**

| Screen Size-mm | Seed recovery % | 1000 seed wt-g | Physical purity-% | Germ-% | SVI |
|---|---|---|---|---|---|
| Un-P | | | | | |
| 1.85 | | | | | |
| 2.50 | | | | | |
| 2.80 | | | | | |
| 4.00 | | | | | |

**Variety: 4**

| Screen Size-mm | Seed recovery % | 1000 seed wt-g | Physical purity-% | Germ-% | SVI |
|---|---|---|---|---|---|
| Un-P | | | | | |
| 1.85 | | | | | |
| 2.50 | | | | | |
| 2.80 | | | | | |
| 4.00 | | | | | |

## Practical – 11

# Importance of Moisture in Seed Processing and Seed Quality Enhancement by Use of Spiral Graders in Soya Bean

The use of seed processing equipments helps in improving the seed quality by inducing uniformity of physical properties of seeds. This help in achieving optimum plant population and desired economic yield. In certain crops like soya bean, standard seed processing may not give desired seed quality because of sensitivity towards moisture. As general practice seed need to be dried to its safe moisture content of 9 before processing. But in case of soya bean lower seed moisture may result higher mechanical damage after standard seed processing. Drier seed are more vulnerable for mechanical damage. In this practical w try to explore the importance of moisture in seed processing and also need of alternative processing equipments like spiral graders to minimize mechanical damage.

### Expt-1: Role of seed moisture in grading of soya bean

***Procedure***

- Take 100 kg of bulk seed and test the moisture before start of the grading.
- Test the moisture and increase or decrease the seed moisture
- To increase seed moisture: keep the seeds in mesh bags inside a walk in germinator with a humidifier or in a mist chamber
- To decrease seed moisture: sun drying for 2-4 hours is recommended.
- Then grade the seeds with different seed moisture content and test the sample for various seed quality parameters such as germination %, SVI, 100 seed weight-g and physical purity

**Table 1:** Effect of seed moisture in grading of soya bean

| Moisture % | Germ. % | Pure Seed % | SVI | 100 seed weight-g | Bulk seed (Control) |
|---|---|---|---|---|---|
| 8 % | | | | | |
| 9 % | | | | | |
| 10 % | | | | | |
| 11 % | | | | | |
| 12 % | | | | | |
| 13% | | | | | |

## Expt-1: Comparison of seed separation with graders and spiral graders

### *Procedure*

- Take 100 kg of soya bean seed and grade with conventional grader and with a spiral grader or both
- Test the seed samples for various parameters
- Measure the mechanical injury % by soaking 100 seeds in 1% Sodium-hypochlorite for 15 minutes (Refer Annexure- 2 for preparation)
- The intact seeds after test are considered as intact or un-damaged
- The seeds which have swollen and have cracked seed coats should be considered as damaged. Work out percentages.

| Method | Germ % | SVI | Pure Seed % | Mech.Injury % |
|---|---|---|---|---|
| Un proc. | | | | |
| Grader | | | | |
| Spiral | | | | |
| Grader + spiral | | | | |

### Study Questions

List out at least 5 reviews on effect of seed moisture on quality and interpret the results

# Practical – 12

# Seed Drying

**Objective:** *To study the influence of dried on seed quality*

Drying is an important unit operation in seed processing in order to maintain seed ability and vigor . it is advocated that seeds should be harvested soon after they attain physiological maturity to obtain better quality seeds and to minimize field losses. The traditional method of sun drying is inadequate, uncontrolled and time bound, which may not meet the high demand of quality seeds in time. Hence artificial drying, the effect of certain drying parameters like seed moisture content, drying air temperature, airflow rate, depth of seed loading and time of exposure on drying characteristics and viability of seeds are to be determined for safe and economical drying of seeds.

## The important factors to be considered in the design of heated air grain driers

### I Dryer factors

a) Size, shape and type of drier - Thickens of grain useful in designing the holding capacity of dried or amount of grain to be dried per unit time.

b) Seed feeding rate

c) Total drying time

d) Air flow pattern and air distribution system

e) Depth of seed bed in the dryer

f) System of cooling seed (if any)

### II Air factors

a) Velocity and air flow rate per unit mass of the grain, i.e, Air flow pattern and air distribution : Cross flow.

b) Temperatures and relative humidity of the heated air and exhaust air.

c) Static pressure of the air at which it blown and

d) Average ambient conditions

## III Seed factors

a) Type, variety and condition of grain

b) Initial and final moisture content of grain

c) The usage of dried grains

d) Lateral heat of vaporization of grain moisture

e) Sensitivity of species

f) Tolerant temperature species

## IV Heating system

a) Type of fuel and rate of fuel supply

b) Type of burner (for liquid fuel) or

Type of heat exchanger (conduction)

## Activity: Treatments (T)

T1: Seeds without drying (Control)

T2: Seeds dried for half day or 4 hours

T3: Seeds dried for one day or 8 hours

T4: Seeds dried under shade for half day or 4 hours

T5: Seeds dried under shade for one day or 8 hours

T6: Seeds dried under direct sun for one day or 4 hours

T7: Seeds dried under direct sun for one day or 8 hours

Repeat the treatments if safe moisture levels not achieved

## Results

| Treatments | Germination % | SVI |
|---|---|---|
| T1 | | |
| T2 | | |
| T3 | | |
| T4 | | |

| T5 |
|---|
| T6 |
| T7 |

## Study Questions

1) List out different seed dryers and principle of operations

   List out seed dryers for different crops and capacity

# Practical – 13

# Seed Treating Methods, Equipments and Evaluating the Quality of Treated Seeds

**Objective:** To study the seed treatment methods and equipments A key factor in achieving the few potential of good quality seed is the blanket treatment of the seeds with agrochemical in order to control or prevent diseases in the crops.

## Purpose/need of the seed treatment and classification

1. Pelleting/seed coating to change the shape of the seed to improve planting value of the seed. Ex: sugar beet.
2. Protection against pest and diseases

a) **Seed disinfection:** It refers to the eradication of fungal spores that have become established within the seed coat, or in depth of the related tissue.

b) **Seed disinfestations:** Seeds is infestation refers to the destruction of surface borne organisms that have contaminated the seed surface but not infected the seed surface.

c) **Seed protection:** The purpose of seed protraction is to protect the seed and young seedling from organisms in the soil, which might otherwise cause decay of the seed before germination.

3. Treatments to overcome dormancy
4. Enhancement of germination and growth
5. Treatment to counteract deterioration

## Properties of Seed Dressing Chemicals

- It should be effective against all the major pathogenic organisms
- It should be non-toxic to plants and people

- It should be environmentally safe
- It should be stable during storage period
- It should be stable during storage period
- It should be systemic in the plant to increase its effective life
- It should be economically competitive.

## Type of seed treatment

- Physical treatment
- Chemical treatment
- Biological treatment

## Physical treatment

a. Hot water treatment
b. Anaerobic water treatment
c. Dry heat treatment
d. Solar heat treatment
e. Scarification
f. Aerated – steam treatment
g. Irradiation treatment
    i) X-ray treatment
    ii) UV treatment
    iii) Y ray treatment

## Chemical treatment

a) **Steep treatment:** The seeds are soaked in a fungicidal liquid solution or suspension for a definite period of time and then drained and dried.

b) **Sprinkle treatment:** The seeds are mixed with a fungicidal liquid solution or suspension and leave it for certain period of time and then dried.

c) **Slurry treatment:** The seeds are mixed with a dust fungicide is a special treater in which small=calibrated mounts of liquid are added.

d) **Dust treatment :** The seeds are mixed with a dust fungicide is a mixing device until the seed is well coated.

e) **Wet treatment:** The seeds are mixed with relatively small amount of concentrated liquid.

There is no recovery of liquid after treatment, but drying is necessary.

a) **Micronutrients:** where soil is known to be deficient in a particular micro-nutrient treatment of the seed with a simple salt soulation may help to overcome to this.

ii. **Growth promoters:** the idea of adding growth stimulants to seeds to accelerate emergence is very attractive.

iii. **Repellents:** Especially where precision sowing is used to achieve an accurate plant population application of repellent is useful in repelling birds, small rodents and insects.

iv. **Colours :** Coloring agents makes seed look attractive and indicates the identify of particular company. Bright colors make the visible in the soil, which may enable the farmer to check the sowing rate or spacing, particularly when precision sowing technique are being used. Ex., dark blue, tends to make seed unrecognizable to birds and may thus reduce losses prior to emergence.

v) **Seed pelleting :** This is a technique for making irregular seeds round to facilitate mechanical precision sowing and should not beconfused with chemical seed treatment. The quantity of material added during pelleting is much grater than is ever applied with a chemical treatment. The pellet must have sufficient integrity to hold together without crumbling during transport and handling. On the other hand, it must breakdown easily in the soil and must offer no significant obstruction to water uptake or radical emergence.

vii) **Scarification :** Immersion in strong acids such as sulfuric or hydrochloric may have a effect in removing of unwanted appendages from seed.

f) **Fumigation :** The seed is treated in an airtight container for a definite period of time by use of volatile material, a fungicide or insecticide.

**Osmotic priming:** This is effectively a pre-germination treatment in which the seed is steeped in a solution of high osmotic potential. This enables full hydration of the tissue with only limited biochemical activity. When sown, primed seed germinates much more quickly and uniformly, which is of a great value in the intensive production of young plants, particularly of high value vegetable crops. Primed seed may be encapsulated in a pellet to protect it and to facilitate sowing but this requires very precise control and can only be justified for high value seeds.

## Biological treatment

**Biological control:** They are in favour as environment concerns about the use of agrochemicals increase. Ex., Rhizobia, VAM, P solubilizing bacteria etc.

## Seed treating equipments

A good seed treater should permit application of small amounts of chemicals to seed and spread it uniformly over the surface of each seed, seed treatment chemicals, fungicides or insecticides must be uniformly applied to all seed in the lot to be effective.

a) Metering device for seed : Metering a known amount of seed into the treating chamber to apply chemical at specified rates. It is necessary to blend a known amount of the chemical with a known amount of seed.

   Common method of metering seed is use of weighted bucket, which collects seed entering the treater and it dumps the known amount of seed into treating chamber.

   Another method of metering seed into the treating chamber is an adjustable speed screw conveyor.

   The speed of the conveyer can be adjusted to deliver a certain volume of speed within a given time.

b) Metering device of chemical: To control over doses of the chemical, which affect seed germination and also to overcome under doses which may not effective against pest, so a metering device is essential to treating accurate rate/dose.

   Some treaters use small dippers of varying capacity for definite rates of application of chemicals and some small metering buckets also used. Some treater spreys the chemicals (liquid) under pressure. The rate of application can be altered by the pressure, nozzle size or number of nozzles in the treater.

c) Methods of applying treatment to the seed

   Five methods application are in use, depending on the nature of the chemical applied.

   i. Dry or dust treatment : Dry powder is applied

   ii. Slurry treatment : A liquid is mixed with the dry powder to make a soup like suspension or slurry, which is then applied to the seed.

   iii. Liquid treatment : A concentrated liquid solution, usually of a volatile fungicide is applied directly to the seed.

   iv. Spray treatment : The liquid or slurry treatment is sprayed on the seed.

   v. Mist treatment : The liquid or slurry is converted to a mist, seed are then passed through the mist.

d) Mixing chamber: The treated seed mass must be thoroughly mixed and agitated to spread the small dose of treatment over all surfaces of each seed. A common method of improving uniformity of treatment is to pass the treated seed through an auger. The mixing auger may have flat metal agitating vanes resembling the flights of conveying augers. Some mixing augers use brushes to agitate and convey seed. Some treater use revolving drums to mix and spread.

e) Means of discharging treated seed: after treating seed may be discharged into a bin or truck or the treater may be equipped with a bagging attachment to spout seed directly in to bags.

Modern seed processing plants spout treated seed into a bin with an automatic bagger weigher, which will accurately meter the desired mount of seed into the bag.

## Study Questions

1. Design standard protocols of seed treatment for different crops

| Sl no | Crop | Polymer -ml/kg | Water- ml/kg | Fungicide -g/kg | Pesticide -ml/kg | Target disease or pest |
|---|---|---|---|---|---|---|

# Practical – 14

# Effect of Pre-sowing Seed Treatment on Quality of Seedlings

Seed is the primary and essential starting point of a wide range of crops, including the majority of vegetables except those vegetative propagated (e.g- potato and sweet potato). Seedling emergence and field stand establishment is one of the problems facing the growers, especially in early planting where adverse conditions are prevailed (low temperature and high soil moisture). Seed germination and seedling emergence are affected by many factors such as: seed genotype, seed quality (seed viability and seed vigor) and the environmental conditions (moisture and temperature) prevailing during seed germination and subsequent seedling emergence, assuming that the seed is not dormant.

Delayed, erratic germination and emergence, poor stand, slow early seedling growth rate and non-uniform maturity are often limit crop production even under optimum environmental conditions Also this wide spread pre-damping off mortality caused by soil born fungi allows the weeds to be well established before the seedlings crop are large enough to be cultivated, compete with the main crop and moreover, they interrupt the cultural practices such as fertilization, chemical application and mechanical harvesting. This experiment tries to establish importance of pre-sowing seed treatment on quality of seedlings in case of hot pepper

## Procedure

- Take 3 seed samples of hot pepper in two sets
- Treat one seed sample with VITAVAX (0.2 % Carboxin 17 % + thiram 17 %) and other with distilled water as control.
- Wet basis also seeds can be treated by putting 100 seeds in a funnel and adding 25ml of VITAVAX solution (for <one minute)
- If the seeds are treated on DRY basis directly sow the seeds in 98 plug plastic trays.

- If the seeds are treated on WET basis dry them and sow the seeds in trays as above
- Watering from time to time and temperature should be maintained for ease in seed germination.
- At the termination of the experiment, 35 days after planting, total survival seedlings, normal and abnormal seedlings according to the ISTA regulation were recorded.

| Treatment | SE% | NS% | ABS% |
|---|---|---|---|
| Control | | | |
| VITAVAX | | | |

SE%: Seedling emergence %
NS%: Normal seedling %
ABS%: Abnormal seedling %

## Study Questions

a) Collect at least one research paper on pre-sowing seed treatment which enhances the seed quality during the entire crop growth period.

## Pre-sowing seed treatments

1) Seed hardening
2) Seed Pelleting
3) Seed coating
4) Seed fortification
5) Seed infusion
6) Seed priming

b) Write interpretation of these research paper highlighting their role in seed quality enhancement

# Practical – 15

# Seed Priming

Seed Priming is the process of controlled hydration of seeds to a level that permits pre-germinative metabolic actively to proceed, but prevents actual emergence of the radical. It is done to increase the speed and uniformity of germination particularly under adverse conditions of temperature, moisture and salinity conditions. In this condition, seeds go through all the essential preparatory processes of germination but prevent cell elongation and radical emergence. During the waiting period at the osmotic barrier the 'slower' seeds tend to catch up with the 'faster ones' so that the subsequent germination is more uniform, than that of untreated seeds. The duration of priming varies with the crop as well as method of priming such that priming is allowed for I and II stage of seed imbibitions without allowing radical protrusion.

Different terms are used depending upon the method adopted for priming.

1. **Osmo-priming**: soaking the seeds in osmotic solution.
2. **Halo-priming**: soaking the seeds in salt solution.
3. **Bio-priming**: coating the seeds with biological agents like bacteria and soaking in warm water.
4. **Solid matrix priming**: This consists of mixing the seeds with an organic or inorganic carrier and water for a period of time. The moisture content of the matrix is brought to a level just below what is required for radicle protrusion. During priming, the water is largely held by the carrier. Seeds can imbibe water from the carrier only till the equilibrium is reached.
5. **Hydro priming**: soaking in excess of water. It is synonymous to hardening with water. Water movement into the seed is not controlled but germination of the seeds is prevented by anaerobic condition.
6. **Drum priming**: subjecting the seeds to water mist while rotating the seeds in drum without allowing radicle protrusion.

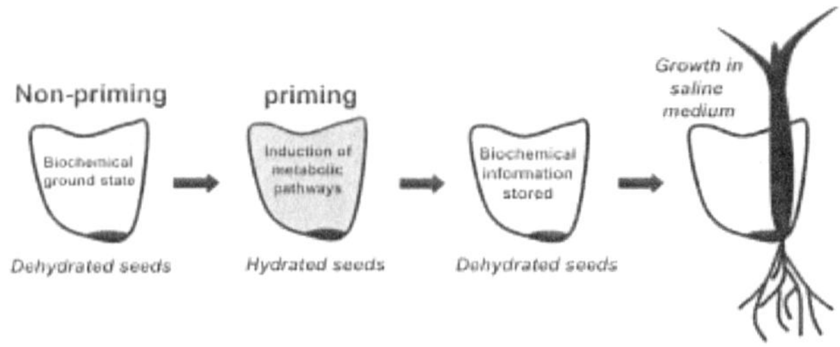


**Fig.1:** Illustration of seed priming procedure

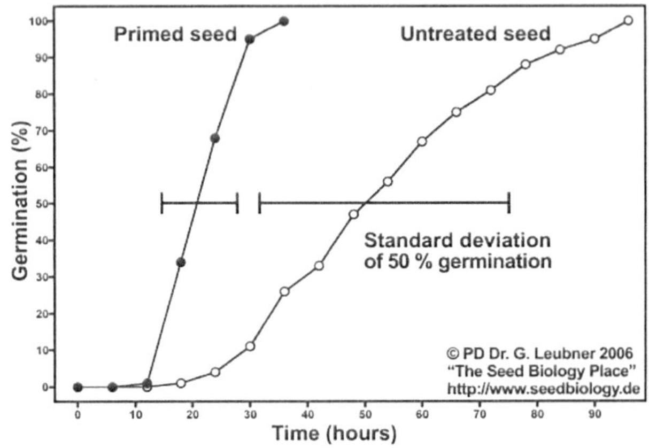


**Fig. 2:** Effect of priming in treated and un-treated seeds

## Activity: OSMOPRIMING

### Materials

- Beaker
- Poly ethylene glycol (PEG 6000) (-1.0 Moa solution)-(Dissolve 273 g in one litre of water)
- Seeds of tomato/brinjal/chilli
- Germination paper

### Procedure

1. Take PEG 6000 (-1.0 MPa) solution in the beaker.
2. Soak the seeds in PEG solution for 3 days.
3. Decant PEG, wash the seeds in water thoroughly and shade dry them to original moisture content.

4. Subject the treated seed to germination test along with the control (untreated) seed.
5. Observe the speed of germination, seed germination and seedling vigor.
6. Record the difference between primed and control seeds.

## Observations

**Table 1:** Effect of seed priming on Germ & RoG

| T* | Germination% | | | | RoG* | | | |
|---|---|---|---|---|---|---|---|---|
| | R-1 | R-2 | R-3 | M # | R-1 | R-2 | R-3 | M |
| Control | | | | | | | | |
| Primed | | | | | | | | |

* Rate of seed germination ( R ) was calculated according to Ellis and Roberts, (1980) = "n /" Dn

Where $n$ is the number of seeds germinated on day $D$, $D$ is the number of days counted from the beginning of the test and $R$ is mean germination rate.

**Table 2:** Effect of seed priming on SVI & SDW

| T* | SVI | | | | SDW** | | | |
|---|---|---|---|---|---|---|---|---|
| | R-1 | R-2 | R-3 | M # | R-1 | R-2 | R-3 | M |
| Control | | | | | | | | |
| Primed | | | | | | | | |

**At the end of germination test (21 days), radicles and shoots were cut from the cotyledons and then dried in an oven at 75±2°C for 24 hours. The dried radicles and shoots were weighted to the nearest milligram and the mean radicle and shoot dry weight and consequently mean **seedling dry weight** were determined.

## Study Questions

Explain the mechanism involved in various types of seed priming and how these influence seed quality at cellular level-Interpret with the help of research articles.

# Practical – 16

# Designing a Low Cost Drum Seed Priming Unit

## Objective

1. Design drum priming (2-3 kg and drying unit with house hold or commonly available materials.
2. Standardization of time for seed priming for different crops.

## Probable drum priming unit components

A] A vertical drum which need to be rotating

B] A motor which will rotate

C] A desktop assembly

D] A simple two way seed drier (Fig not given here)

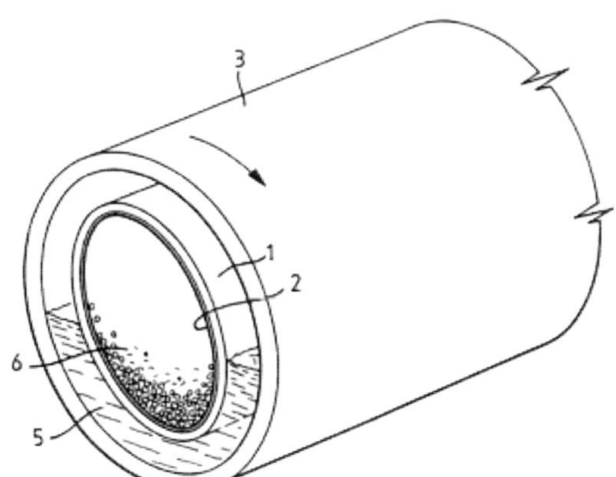


**Fig. 1.**

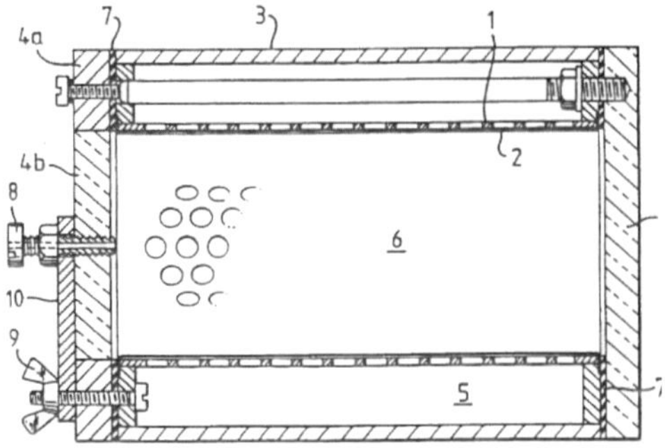


Fig. 2.

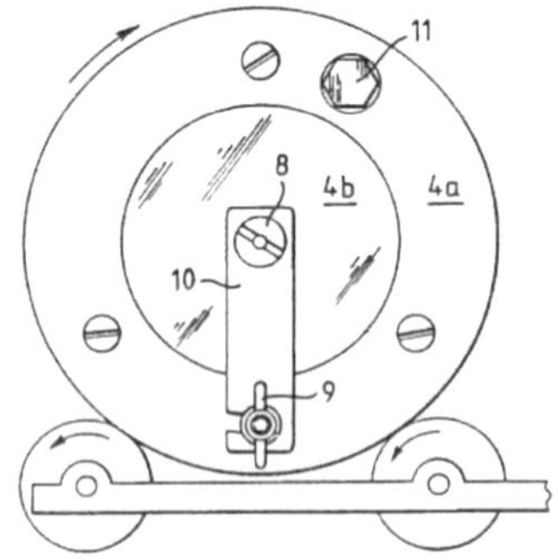


Fig. 3.

## Working principle

1. Seed priming is a process for treating plant seeds that enables them to undergo faster and more uniform germination on sowing or planting
2. This process allows the seeds to absorb enough water to enable their pre-germinative metabolic processes to begin and then arrests them at that stage.
3. The amount of water absorbed must be carefully controlled as too much would simply allow the seed to germinate and too little would result in the seed ageing.
4. Once the correct amount of water has been absorbed it is then necessary to hold the seed at that water content for a period, typically one to two weeks, before drying it back to the original water content for storage.

Table for calculations

| Crop | Germination % | | |
|---|---|---|---|
| | Natural Seed | Primed-Fresh | Primed- Dry |
| | | | |

## Study Questions

List the time taken (days) for germination in different crops under natural, primed-fresh and primed-dry conditions

# Practical – 17

# Seed Pelleting

The objective of seed pelleting is to provide an opportunity to package effective quantities of materials such that they can influence the micro-environment of each seed.

## Process of Pelleting

Seeds are introduced into a coating drum or pan that resembles a cement mixer. An amalgam of pelleting materials (clays, limestone, calcium carbonate, talc, vermiculite) and cementing adhesives (gum Arabic, gelatine, methyl cellulose polyvinyl alchol) are used to form the pellet and other compounds such as inoculants, fungicides, etc. May be added to enhance seed performance.

As drum rotates, the seeds are first sprayed with water followed by the addition of the pelleting materials with binder. The wet seed attracts and becomes coated with the dry pelleting material and the pellet gradually increases in size with each turn of coating drum. Longer rotation times with greater amounts of pelleting materials lead to greater pellet size and roundness. At the end of the pelleting process, a binder is added to harden the outer layer of the pellet. This also reduces the amount of dust produced during handling, shipping and planting. After pelleting is complete, the pelleted seed are dried and handled in the same way as un-pelleted seeds.

**Fig. 1:** Seed Pelleting in vegetables

## Activity: Pelleting of Seeds

### Materials

- Plastic tray
- Cooled solution of 1% Maida
- Leaf powder of *Albizza amara or neem*
- Dia-mmonium phosphate
- Onion seeds

### Procedure

1. Take the seeds in a plastic tray
2. Boil 100 ml of water and add 1 g of maida and stir vigourously. Allow the solution to cool.
3. Add a small quantity of cool maida solution (1%) to the seeds.
4. Shake vigourously, so that the adhesive spreads evenly on all the seeds.
5. Add pelleting material (Diammonium phosphate + leaf powder of *Albizza amara* or neem evenly over the seed and continue shaking. The wet seed surface will attract the pelleting material and result in even coating over the seeds.
6. Separate the seed clumps formed, if any, manually.
7. Empty the seeds on a sheet of paper and allow it to dry for a day.
8. Subject the pelleted seed to germination test along with control.
9. Observe the speed of germination, germination percent and seedling vigour.

**Table 1:** Effect of seed pelleting on Germ & RoG

| T | Germination% | | | | RoG* | | | |
|---|---|---|---|---|---|---|---|---|
| | R-1 | R-2 | R-3 | M" | R-1 | R-2 | R-3 | M |
| Control | | | | | | | | |
| pelleted | | | | | | | | |

* Rate of seed germination ( R ) was calculated according to Ellis and Roberts, (1980) $= \Sigma n / \Sigma Dn$

Where *n* is the number of seeds germinated on day *D*, *D* is the number of days counted from the beginning of the test and *R* is mean germination rate.

**Table 2:** Effect of seed pelleting on SVI & SDW

| T* | SVI | | | | SDW** | | | |
|---|---|---|---|---|---|---|---|---|
| | R-1 | R-2 | R-3 | M# | R-1 | R-2 | R-3 | M |
| Control | | | | | | | | |
| pelleted | | | | | | | | |

** At the end of germination test (21 days), radicles and shoots were cut from the cotyledons and then dried in an oven at 75±2°C for 24 hours. The dried radicles and shoots were weighted to the nearest milligram and the mean radicle and shoot dry weight and consequently mean **seedling dry weight** were determined.

## Study Questions

1. List the materials used in seed Pelleting
2. What are different types of seed Pelleting.
3. Explain procedure involved in sampling, coating and rolling

# Practical – 18

# Study of Packaging Materials

These materials include barrier coated paper bags, barrier laminated film bags, jute bags, PP woven box bag and cotton bags.

## A: Paper Bags

1. **Coated paper**: bags provide a barrier for leaking or staining of the product. This can be done by a greaseproof or claycoated kraftpaper.
2. **Open-Mouth sacks**: can be manufactured using different types of paper. Simple kraftliner, glossy claycoated and lacquered outerliner are possible. These sacks can be supplied with liners and printed up to 8 colours.
3. **Pinch-bottom paper bags**: have folded sides. The bags can be provided with a hot melt top to facilitate closing. Special features include: embossing, valve dimensions and ultrasonic valve closure. Artwork can be printed up to 8 colors.
4. **Block bottom bags**: can be manufactured with handles, allocated in either top, gusset or bottom. Options like a net or PE based window, allows a clear presentation of the product content.
5. **Block bottom bags**: are available in different sizes and specifications to fit a content of 500 grams to 10 kg. Crossbottom bags for food and non-food can fit a content of up to 25 kg product. Paperquality, perforation types and special lacquers can be applied to allow for the best protection of the packaging content.
6. **UV-Lac Bags**: These bags can be provided with windows, UV-lacquering, bottomhandles and easy-opening systems.
7. **Premade bags**: made from a combination of laminate film and paper. Paper-fresh film is suited for automatic filling on vertical form fill and seal machines.

## B. Film Packaging

1. **Barrier bags/Barrier film bags:** multilayer laminate bags which protect the packaging content against degradation by oxygen, gas and sunlight. Using a customised barrier layer, including aluminium or metalised polyester, will benefit the shelflife of several products.
2. **Film sheets:** PE and PP film sheets for covering crates and boxes of fresh producethat enhances the ventilation process.
3. **Flow-Fresh bags:** is a combination of PP film and PP runs on standard flowpack machines and horizontal or vertical form, fill and seal machines.
4. **Sachets**: for seeds. Sachets are a flexible, convenient, attractive and economic packaging solution. These sachets are made from high barrier laminate materials to protect the packaging content against the influence of oxygen, gas and light.

Coated paper bags

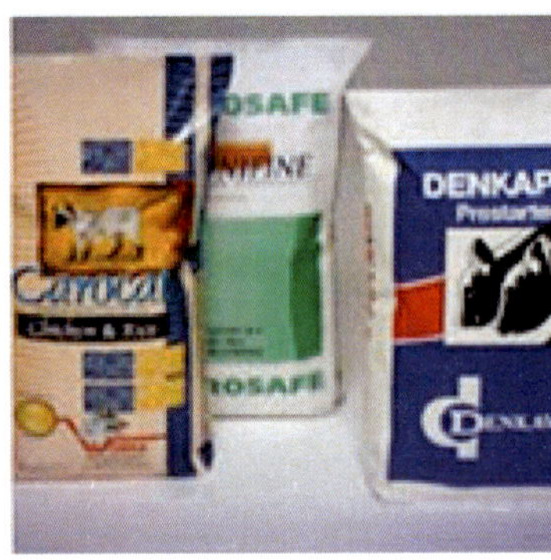

Open-Mouth paper sacks

Paper bags without handle

SOS Blockbottom paper bags

Pinch-bottom paper bags

Paper bags with handle

Laminated paper bags

Barrier bags

Barrier film

Film sheets

PP, PE bags

PP, PE flowpack

PVC, non, PVC stretchfilm

Flow-Fresh® (PP film + PP CLAF net)

Standup pouch (Dovpack)

Sachets

Twin-Bag®(PE film+PE CLAF net)

Jute bags for potatoes

Jute bags for seeds

Jute bags for mussles

Jute bags for retail

## C. Jute Bags

Jute is one of the oldest traditional packaging materials. It was the most commonly used packaging material. This purely natural material is made from the bark of the jute plant. It is strong and durable and has ventilating and absorbent properties. The structure of the fabric prevents damage to the packaged product and so it is still frequently used today due to its excellent qualities. The jute bags can be supplied in various sizes to fit a content upto to 50 kg. Jute is used for transport and retail packaging. The bags can be provided with a wicket and a drawstring.

1. **Woven Bags:** PP woven bags are ideally suited for applications where strong and cost efficient packaging is requested. In emergency situations, such as floodings, we can supply woven polypropylene bags to be used as sandbags.The PP box bag with mosquito fabric is specially designed to allows drying after and during storage.
2. **Box Bag:** The NNZ box bag has been developed for freshly harvested seeds and grains. The fine gitter bottom allows good ventilation, so the (grass)seeds and grains can be dried effective.

PP wovenbag

PP shopping bag

## D. Alumium Tins/ Foils

These types of packaging use aluminum foil up to thickness of 6-200 microns. Usually done for high value seeds and this type packaging useful in storing seeds for longer duration and also help in production, storage, distribution and retailing.

## Activity

### Treatments

### Factor-1: Initial Seed Moisture (M)

M1: 8 %

M2: 10%

M3: 12%

### Factor-2: Storage period (S)

S1: 0 month

S2: 1 month

S3: 2 month

S4: 3 month

S5: 4 Month

### Factor-3: Packaging Material (P)

P1: Gunny bag (GB)

P2: Poly lined bags(PB)

P3: Al-foils(AF)

P4: Plastic containers (PC)

## Observations

1. Germination%

| Initial Moisture Content %( M) | Storage Period-m (S) | Germination % Packaging Material (P) | | | |
|---|---|---|---|---|---|
| | | GB | PB | AF | PC |
| 8% | 0 | | | | |
| | 1 | | | | |
| | 2 | | | | |
| | 3 | | | | |
| | 4 | | | | |
| 10% | 0 | | | | |
| | 1 | | | | |
| | 2 | | | | |
| | 3 | | | | |

| | | | | | |
|---|---|---|---|---|---|
| | 4 | | | | |
| 12% | 0 | | | | |
| | 1 | | | | |
| | 2 | | | | |
| | 3 | | | | |
| | 4 | | | | |

2. Seedling vigour index (SVI)

| Initial Moisture Content %( M) | Storage Period-m (S) | SVI Packaging Material (P) | | | |
|---|---|---|---|---|---|
| | | GB | PB | AF | PC |
| 8% | 0 | | | | |
| | 1 | | | | |
| | 2 | | | | |
| | 3 | | | | |
| | 4 | | | | |
| 10% | 0 | | | | |
| | 1 | | | | |
| | 2 | | | | |
| | 3 | | | | |
| | 4 | | | | |
| 12% | 0 | | | | |
| | 1 | | | | |
| | 2 | | | | |
| | 3 | | | | |
| | 4 | | | | |

## Study Questions

Work out cost of packaging for at least one crop each in cereals, pulses, oil seeds, vegetables, fruits for the following types

1. Cloth bag
2. Al-foils
3. Sachets

## Practical – 19

# Seed Fumigation

Fumigation is a method of pest control that completely fills an area with gaseous pesticides to suffocate or poison the pests within. It is utilized for control of pests in buildings, soil, seed, and grain produce. It is also used during processing of goods to be imported or exported in order to prevent transfer of exotic organisms. Fumigants are chemical compounds in gaseous form that enter the body of insect through the spiracles spread all over via trachea and tracheoles. The important fumigants are Methyl bromide, Ethylene dibromide, Aluminium phosphide, Ethyl formate, Hydrogen cyanide, Chloropicrin etc.

Phosphine fumigant protected wheat seeds against the infestation of lesser grain borer *Rhyzopertha dominica* whereas, 7.33 per cent infestation was observed in unfumigated seeds after four months of storage. Further there was no adverse effect of repeated fumigations at three folds higher dose (9 g/m$^3$) on seed germination. (Gupta *et al.*, 2000)

Singh *et al.* (2003) reported that green gram and chick pea seeds (having 12 per cent moisture content) fumigated with methyl bromide and stored at ambient temperature, lost their germination within 24 months, while the unfumigated seed in same condition retained 97 per cent germination. Whereas, at 9 per cent seed moisture, methyl bromide fumigated seeds maintained viability up to 42 months but in phosphine fumigated seeds started to decline in germination after 21 months and were completely dead by 37 months.

Groundnut seeds fumigated with ethylene dibromide for 24 hour recorded 88 per cent germination, seedling length (16 cm) and seedling vigour (1355). Whereas, in unfumigated seeds germination per cent, seedling length and vigour index was 82, 12, and 1014 respectively. Further during ten months storage germination per cent, seedling length and vigour index declined to 58, 10, and 635 respectively. (Vijayanna, 2006).

Although fumigation is one of the most effective method to control the storage insects, but it is effective only when seeds are fumigated with desired fumigant at desired quantity at normal temperature and moisture content with minimum exposure period. Among different fumigants, aluminium phosphide is considered as one of the safest fumigant as it is not having deleterious effect on seed quality.

### Al-Phosphide: Dosage: 9 gm/T

### Packing size: 3 to 12 gm

**Safe frequency:** For Cereals and pulses -2 times and oilseeds-once

### Precautions

- Fumigation should be the last resort if the inventory is going to be stored for longer time
- The gases produced from Al-P tablets tend to suspend at the top so it is better to insert tablets between stacks at lower ends
- Take appropriate care to sanitize the storage facility. Usually the field carry over and rejected seed are active source of insect eggs or pupa.
- For sanitation spraying once in a month is advisable with Nuvan and Deltamethrin @ 4ml/L is very helpful. Spray on the top, bottom and corners of bags and stacks

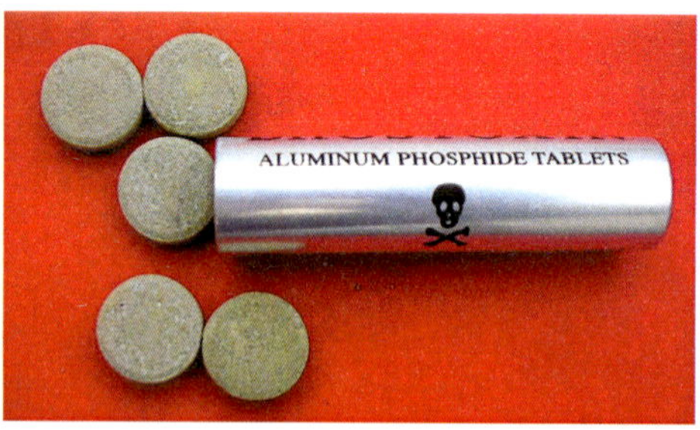


**Fig.1:** Al-Phosphide tubes

**Fig. 2:** Seed fumigation

**Fig. 3:** Seed sanitation

**Table 1:** Effect of seed fumigation on seed quality

| Lot 1 | GERM % | SVI | Pest incidence % |
|---|---|---|---|
| No Fumigation | | | |
| Sanitation only | | | |
| Fumigation only | | | |
| Sanitation+Fumigation | | | |
| Lot 2 | GERM % | SVI | Pest incidence % |
| No Fumigation | | | |
| Sanitation only | | | |
| Fumigation only | | | |
| Sanitation+Fumigation | | | |
| Lot 3 | GERM % | SVI | Pest incidence % |
| No Fumigation | | | |
| Sanitation only | | | |
| Fumigation only | | | |
| Sanitation+Fumigation | | | |
| Lot 4 | GERM % | SVI | Pest incidence % |
| No Fumigation | | | |
| Sanitation only | | | |
| Fumigation only | | | |
| Sanitation+Fumigation | | | |
| Lot 5 | GERM % | SVI | Pest incidence % |
| No Fumigation | | | |
| Sanitation only | | | |
| Fumigation only | | | |
| Sanitation+Fumigation | | | |

## Study Questions

1) List out various types fumigation permitted in India under plant quarantine
2) Explain the mechanism and mode of action of seed fumigation

# Practical – 20

# Measurement of Seed Moisture Content

The moisture content is the amount of water in the seed and is usually expressed as a percentage. It can he expressed on either a wet weight basis (where it is expressed as a percentage of the fresh weight of the seed) or on a dry weight basis (where it is expressed as a percentage of the dry weight of the seed). The seed moisture content is the most vital parameter, which influence the seed quality and storage life of the seed. Seed moisture content is closely associated with several aspects of physiological seed quality. For example, it is related to seed maturity, optimum harvest time, mechanical damage, economics of artificial seed drying, seed longevity and insect & pathogen infestation.

**Principle :**The methods prescribed are designed to reduce oxidation, decomposition or the loss of other volatile substances while ensuring the removal of as much moisture as possible.

## Determination of seed moisture content

As seed moisture and its management influences so many physiological seed quality parameters essential to seed quality. Seed moisture measurement appropriate to the purpose is needed in commerce and research. The optimum method for moisture testing depends upon:

1. Chemical composition of seed
2. Seed structure
3. Moisture content level
4. Degree of accuracy and precision required
5. Constraints of time
6. Technical expertise and cost

**Direct method :**Under this category, the seed moisture content is measured directly by loss or gain in seed weight. These are:

1. Desiccation method
2. Phosphorus pentaoxide method
3. Oven-drying method
4. Vacuum drying method
5. Distillation method
6. Karl Fisher's method
7. Direct weighing balance
8. Microwave oven method

**Indirect method:** These are no so accurate; estimation is approximate, but convenient and quick in use. These are frequently used at seed processing plants. These measure other physical parameters like electrical conductivity or electrical resistance of the moisture present in the seed. Values are measured with the help of seed moisture meters, and these values are transformed into seed moisture content with the help of calibration charts, for each species, against standard air-oven method or basic reference method.

Above all Karl-Fisher's method has been considered as the most accurate and the basic reference method for standardizing other methods of seed moisture determination. The constant temperature oven drying method is the only practical method, approved by International Seed Testing Association (ISTA) and other organization to be used for routine seed moisture determination in a seed-testing laboratory.

## Constant temperature oven drying method

The constant temperature oven drying method is broadly grouped into two categories:

1) Low Constant Temperature Oven Method
2) High Constant Temperature Oven Method

**Low constant temperature oven method:** This method has been recommended for seed of the species rich in oil content or volatile substances (Table I). In this method, the pre- weighed moisture bottles along with seed material are placed in an oven maintaining a temperature of 103°C. Seeds are dried at this temperature for 17+-1hr.The relative humidity of the ambient air in the laboratory must be less than 70 percent when the moisture determination is carried out.

**High constant temperature oven method:** The procedure is the same as above except that the oven is maintained at a temperature of 130°-133°C. The sample is dried to a period of four hours for Zea Mays, two hours for other cereals and one hour for other species (see alsoTable2).In this method there is no special requirement pertaining to the relative humidity of the ambient air in the laboratory during moisture determination.

## Period of seed drying

The prescribed period of seed drying shall be17-+ 1 hrs at 103°C under low constant and 1 to 4 hrs at 130°-133°C under high constant temperatures. Maize seed be dried for 4 hrs, cereals and/or other millets for 2 hrs and the remaining species for 1 hr. Seeds rich in oil content or with volatile substances be dried for 17+- hrs under low constant temperature.. See drying period begins from the time oven returns to maintain the desired temperatures

**Sample size**: The ISTA rules recommend that two replicates, each with 4 gm of seed be used for determination of seed moisture content.

**Calculation of results:** The moisture content as a percentage by weight (fresh weight basis) is calculated to one decimal place, by using of the formulae:

$$\text{\% age seed m.c} = \frac{M2-M3}{M2-M1} \times 100$$

**Where**

M1 = Weight of the weighing bottle/container with cover in gm

M2 = Weight of the weighing bottle/container with cover and seeds before drying

M3 = Weight of the weighing bottle/container with cover and seeds after drying

Note: The seed moisture determination must be done in two replicates, with precise weighing (i.e. up to three decimal places) using lightweight weighing bottles/containers.

If the seed is pre-dried or dried in two steps: The seed moisture content is calculated from the results obtained in the first (pre-dried) and second stages of seed drying, using the following formula, and expressed as percentages, as under:

$$\text{\%age seed m.c} = \frac{(S1 + S2) - (SI * S2)}{100}$$

Where

S1 = is the moisture loss in the first stage, and

S2 = is the moisture loss in the second stage .

**Reporting of results:** Seed moisture content be reported to the nearest 0.1% on analysis certificate.

| Sample | Repln | Initial Wt-g | Dry wt.-g | % M.C(Dry) | % M.C(Wet) |
|---|---|---|---|---|---|
| | | | | | |

## Study Questions

List out latest equipments for measuring seed moisture with principle involved

# Practical – 21

# The Relationship of Seed Moisture Content and Mechanical Damage

Mechanical damage in oil seeds like soya bean can be induced at any time during harvesting, handling, conditioning and sowing operations. There is relationship between moisture and mechanical damage. In general the low moisture containing seeds are more prone for mechanical damage than the high moisture containing seeds. In this experiment we look for this relationship by measuring seed quality parameters like germination and leakage (electrical conductivity).

## Procedure

Soya bean seeds of varieties JS-335 and DSB-21 were used in this study. The moisture content of seed samples were adjusted to 8 or 12 percent (wet weight basis) and kept in plastic zip lock bags for subsequent studies Mechanical damage was experimentally induced by dropping seeds onto a steel plate from a height of 2.0 m four and eight times times. The seed quality parameters like germination and electrical conductivity were measured following ISTA procedure.

## Data collection and analysis

Collect the data and plot the line or scatter graphs to see the relationships.

**Table 1:** Effect on E. C

| Particulars | Electrical Conductivity μA | | | |
|---|---|---|---|---|
| | R-1 | R-2 | R-3 | Mean |
| **Varieties (V)** | | | | |
| VI: JS-335 | | | | |
| V2: DSB-21 | | | | |
| **No. of Impactions (I)** | | | | |
| I1: 4 times | | | | |
| I2: 8 times | | | | |
| **Moisture (M)** | | | | |
| M1: 8% | | | | |
| M2: 9% | | | | |
| M3: 10% | | | | |
| M4: 11% | | | | |
| M5: 12% | | | | |

**Table 2:** Effect on Germination %

| Particulars | Germination % | | | |
|---|---|---|---|---|
| | R-1 | R-2 | R-3 | Mean |
| **Varieties (V)** | | | | |
| VI: JS-335 | | | | |
| V2: DSB-21 | | | | |
| No. of Impactions (I) | | | | |
| I1: 4 times | | | | |
| I2: 8 times | | | | |
| **Moisture (M)** | | | | |
| M1: 8% | | | | |
| M2: 9% | | | | |
| M3: 10% | | | | |
| M4: 11% | | | | |
| M5: 12% | | | | |

## Study Questions

List out faster detection methods of seed mechanical damage in various crops.

# Practical – 22

# Seed Storage : Validating Harrington's Rules for a Particular Area

According to Harrington's rule (1960) of thumb, for storage, every one percent increase in moisture content, the seed life is halved far seeds of moisture content of 5-14%. For temperature, every 5°C rise in temperature, between 0-50 °C, the seed life is halved.

### In summary Harrington's rule states that:

1. Each 1 percent reduction in moisture content doubles the life of the seed.
2. Each 10 °F reduction in temperature doubles the life of the seed

The purpose of this experiment is to know whether these rules are area specific which may vary with temperatures and humidity.

### Reference

Harrington, J.F. (1960). Thumb rules of drying seed. *Crops & soils*, **13**, 16-17

### Activity

1) Take sets of seven seed samples of various seed moisture from 15 to 9%, keep them for storage for various period (1-6 months) and test for germination %
2) Interpret the results based on the reduction in moisture and improvement in germination
3) Record the observations as below in the tables provided (Table-1)
4) The second set of seven samples used to keep in controlled storage at various temperatures
5) Record the observations in Table-2

**Table-1:** Effect of reduction in seed moisture
Initial Seed Moisture: 15%

Crop:
Storage period: one month

| Seed Moisture % | Germination % | | | |
|---|---|---|---|---|
| | R-1 | R-2 | R-3 | Mean |
| 15 | | | | |
| 14 | | | | |
| 13 | | | | |
| 12 | | | | |
| 11 | | | | |
| 10 | | | | |
| 9 | | | | |

Initial Seed Moisture: 15%
Crop:
Storage period: 2 months

| Seed Moisture % | Germination % | | | |
|---|---|---|---|---|
| | R-1 | R-2 | R-3 | Mean |
| 15 | | | | |
| 14 | | | | |
| 13 | | | | |
| 12 | | | | |
| 11 | | | | |
| 10 | | | | |
| 9 | | | | |

Initial Seed Moisture: 15%
Crop:
Storage period: 3 months

| Seed Moisture % | Germination % | | | |
|---|---|---|---|---|
| | R-1 | R-2 | R 3 | Mean |
| 15 | | | | |
| 14 | | | | |
| 13 | | | | |
| 12 | | | | |
| 11 | | | | |
| 10 | | | | |
| 9 | | | | |

Initial Seed Moisture: 15%
Crop:
Storage period: 4 months

| Seed Moisture % | Germination % | | | |
|---|---|---|---|---|
| | R-1 | R-2 | R-3 | Mean |
| 15 | | | | |
| 14 | | | | |
| 13 | | | | |
| 12 | | | | |
| 11 | | | | |
| 10 | | | | |
| 9 | | | | |

Initial Seed Moisture: 15%
Crop:
Storage period: 5 months

| Seed Moisture % | Germination % | | | |
|---|---|---|---|---|
| | R-1 | R-2 | R-3 | Mean |
| 15 | | | | |
| 14 | | | | |
| 13 | | | | |
| 12 | | | | |
| 11 | | | | |
| 10 | | | | |
| 9 | | | | |

Initial Seed Moisture: 15%
Crop:
Storage period: 6 months

| Seed Moisture % | Germination % | | | |
|---|---|---|---|---|
| | R-1 | R-2 | R-3 | Mean |
| 15 | | | | |
| 14 | | | | |
| 13 | | | | |
| 12 | | | | |
| 11 | | | | |
| 10 | | | | |
| 9 | | | | |

**Table 2:** Effect of storage temperature

Crop:

Seed Moisture: 15 %

Storage period: 1 month

| Storage Temperature (° F) | Germination % | | | |
|---|---|---|---|---|
| | R-1 | R-2 | R-3 | Mean |
| 90 | | | | |
| 80 | | | | |
| 70 | | | | |
| 60 | | | | |
| 50 | | | | |
| 40 | | | | |
| 30 | | | | |

Crop:

Seed Moisture: 14 %

Storage period: 2 months

| Storage Temperature (° F) | Germination % | | | |
|---|---|---|---|---|
| | R-1 | R-2 | R-3 | Mean |
| 90 | | | | |
| 80 | | | | |
| 70 | | | | |
| 60 | | | | |
| 50 | | | | |
| 40 | | | | |
| 30 | | | | |

Crop:

Seed Moisture: 13 %

Storage period: 3 months

| Storage Temperature (° F) | Germination % | | | |
|---|---|---|---|---|
| | R-1 | R-2 | R-3 | Mean |
| 90 | | | | |
| 80 | | | | |
| 70 | | | | |
| 60 | | | | |
| 50 | | | | |
| 40 | | | | |
| 30 | | | | |

Crop:

Seed Moisture: 12 %

Storage period: 4 months

| Storage Temperature (° F) | Germination % | | | |
|---|---|---|---|---|
| | R-1 | R-2 | R-3 | Mean |
| 90 | | | | |
| 80 | | | | |
| 70 | | | | |
| 60 | | | | |
| 50 | | | | |
| 40 | | | | |
| 30 | | | | |

Crop:
Seed Moisture: 11 %
Storage period: 5 months

| Storage Temperature (° F) | Germination % | | | |
|---|---|---|---|---|
| | R-1 | R-2 | R-3 | Mean |
| 90 | | | | |
| 80 | | | | |
| 70 | | | | |
| 60 | | | | |
| 50 | | | | |
| 40 | | | | |
| 30 | | | | |

Crop:
Seed Moisture: 10 %
Storage period: 6 months

| Storage Temperature (° F) | Germination % | | | |
|---|---|---|---|---|
| | R-1 | R-2 | R-3 | Mean |
| 90 | | | | |
| 80 | | | | |
| 70 | | | | |
| 60 | | | | |
| 50 | | | | |
| 40 | | | | |
| 30 | | | | |

## Study Questions

1) List out the other thumb rules of seed moisture and storage temperature

# Practical – 23

# Quantification of Seed Processing Efficiency (SPE)

Seed processing efficiency (SPE) is one of the important aspects in seed processing. It will help in measuring the potential of the shop floor. It will also help in standardizing activities like screens and improving physical purity. SPE will help to improve the quality of the seed material and help in utilizing the machines at their optimum capacity. SPE may vary among the different crops and even among the varieties.

To quantify SPE we need to pass the test sample (bulk-pre processed) through the target machine and enter the details in a processing register. To quantify the SPE we need to follow these mathematical formulae:

### Seed Processing Efficiency (Pure Seed)

$$\textbf{SPE (\%)} = \frac{G_o}{G_o + (G_r + G_b)} \times 100 \qquad (1)$$

Where,

$G_o$ = Good seed at the outlet (kg)

$G_r$ = Rejected seed at the out let (kg)

$G_b$ = Blown away reject at out let (kg)

### Seed Processing Loss:

$$\textbf{SPL (\%)} = \frac{(G_r + G_b)}{G_i} \times 100 \qquad (2)$$

Where,

$G_r$ = Rejected seed at the out let (kg)

$G_b$ = Blown away reject at out let (kg)

$G_i$ = Total seed fed to the machine at feeder (kg)

## Seed Processing Throughput Capacity

$$\textbf{SPTC (kg/hr)} = \frac{G_i}{T} \qquad (3)$$

Where,

$G_i$ = Total seed fed to the machine at feeder (kg)

T = Time taken to completely process the seed (h)

## Visible seed damage

$$\textbf{VSD}\ \% = \frac{D_S}{D_w} \times 100 \qquad (4)$$

Where,

$D_S$ = Visible seed damage by weight (kg)

$D_w$ = Total seed at the feeder (kg)

**Seed Processing Productivity**:

$$\textbf{SPP (kg/h)} = \frac{60 \times G_i}{T} \qquad (5)$$

Where,

$G_i$ = Total seed fed to the machine at feeder (kg)

T = Time taken to completely process the seed (h)

## Study Questions

1. Work out the PE for the current production in the shop floor
2. List other methods and mathematical model to measure PE

# Practical – 24

# Visit to Commercially Controlled and Uncontrolled Seed Storages

---

## NAME OF THE UNIT:

1. Location of Seed Storage
2. Date of Inspections/visit

## A. Seed Store Building, Construction, etc.

i. Type and condition of floor:

ii. Nature of wall:

iii. Nature of ceiling/roof:

iv. Water seepage/rain water leakage:

v. Rat proofness of the go down:

vi. Safety arrangements:

vii. Facilities for aeration, ventilation, etc.

viii. Facilities for air tightness, etc. for fumigation:

## B. Stacking

i. Whether standardized wooden pallets are used for stacking:

ii. Size of stacks (feet):

Length_____Breadth_______Height:________

iii. Size of packing used in stacking:

iv. Distance between the top of stack and ceiling:

v. Whether stack card is used for each lot:

i. Whether stack card entries are up-to-date:

ii. Whether adequate alley space is provided between stacks:

iii. Distance between wall and stack (feet):

iv. Whether proper ventilation is provided:

v. Approximate height of location of ventilator from ground level (feet):

vi. Whether lot and variety identity are maintained in stacks:

## C. Sanitation and insect control

i. Chemical used for fumigation/spraying

ii. Dosage of the Chemical used:

iii. Whether uniform dosage has been used for the entire quantity:

iv. Interval between fumigation/spraying/(days):

v. General range of moisture of seed at the time of fumigation:

vi. Prevalent environmental condition at the Time of fumigation:

vii. Length of exposure of seed to fumigant (hours):

viii. Kind of packing material used for bagging seed:

## D. Revalidation (Minimum 5 seed lots on random basis):

| Crop | Variety | Lot No. | Date of test | Due date for revalidation | Actual date of revalidation |
|---|---|---|---|---|---|

## E. Quality of seed (Minimum 5 seed lots on random basis):

| Crop | Variety | Lot No. | Moisture % age | Remarks on the physical appearance of seed | Remarks on insect infestation disease |
|---|---|---|---|---|---|

## F. Condemnation/disposal

i. Whether seed meant for condemnation has been segregated:

| Crop | Variety | Lot No. | Qty. (Qtls.) | Reasons stated for not disposing so far. |
|---|---|---|---|---|

## H. General cleanliness of the Store

i. Whether seed bags are clean, intact with seal tags, labels, etc.

ii. Remarks on maintenance of:

   a. Germination Register:

   b. Spraying Register:

iii. Condition of file samples:

iv. Approximate age of the oldest seed in the store for each variety and its quality:

| Crop | Variety | Lot No | App. Age (months) | Germination % age | Date of test |
|---|---|---|---|---|---|

## I Storage of unprocessed seed

(Minimum 5 seed lots on random basis):

i. Whether unprocessed seed is being stored Properly:

ii. Condition of unprocessed seed

| Crop | Variety | Lot Number | Remarks |
|---|---|---|---|

## Study Questions

1. Document all the activities as they happen on a shop floor-step wise

| Steps | Activity | Significance |
|---|---|---|

# Practical – 25

# Study of PPE and ESH in Seed Processing

PPEs are the personal protection equipments which protect the personnel against some hazards in the shop floor. PPEs form important aspects of **Environment** (safe disposal of treated chemicals, dusts coming out of processing), **Safety** (to seed storages) and **Health** (personnel) (ESH) strategy of any shop floor.

Some of the hazards such as:

1. the lungs, eg from breathing in contaminated air or dust during seed cleaning
2. the head and feet, eg from falling materials
3. the eyes, eg from flying dust particles or splashes of corrosive liquids of seed treatment
4. the skin, eg from contact with corrosive materials like seed treatment chemicals
5. the ears, eg from noises

## Some of the basic PPE and ESH tools:

1. **Helmets**: If there is any danger of falling objects or head injury, a hard hat should always be worn.

2. **Goggles / safety glasses**: Safety glasses are usually made from shatter-resistant plastic lenses to protect the eyes from flying materials. Although safety lenses may be constructed from a variety of materials that vary in impact resistance, certain standards suggest that they maintain a minimum 1mm thickness at the thinnest point, regardless of material. If chemicals splash in your face or eyes, flush skin and eyes with water for at least 15 minutes and then get medical attention.

3. **Aprons**: To protect against any spillage of seed treatment chemicals which can cause damage to the skin.

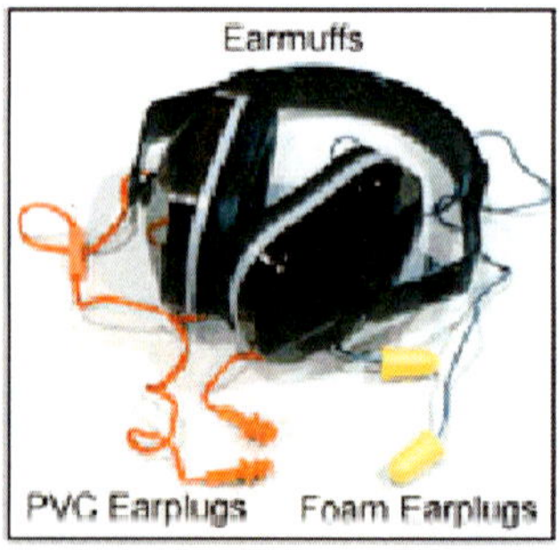


4. **Disposable ear plugs**: Noise is a common problem in many workplaces. High levels of noise can gradually damage your hearing and this is unfortunately a permanent handicap. The following types of hearing protection equipment are available:
   - Foam Earplugs
   - PVC Earplugs
   - Earmuffs

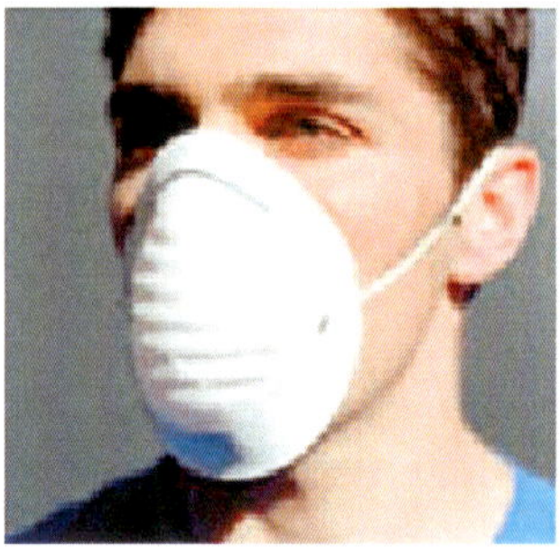

5. **Disposable Masks**:A dust mask should always be worn when working with harmful gases and in areas with a lot of dust. These masks are also to be worn when treating seed with an insecticide and fungicide, especially the powder types.

6. **Gloves**: When working with wooden pallets or any type of material which can cause injury to your hands, protective gloves must be worn. There are many different types of gloves, but you will be provided with those appropriate to your workplace. A special kind of resistant glove is also necessary when working with dangerous chemicals.

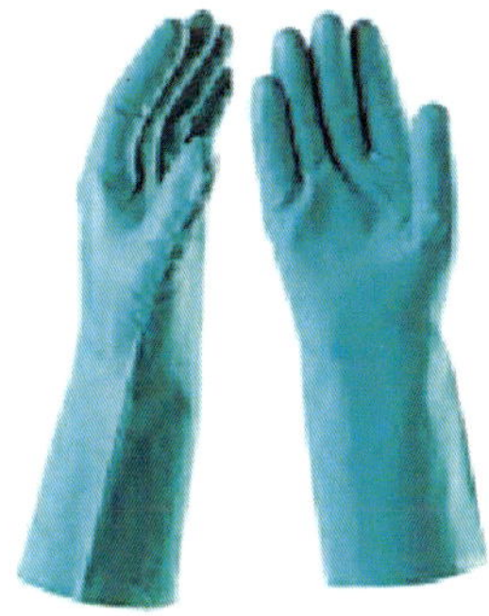

Gloves must be:

- appropriate to the material handled
- worn whenever there is potential for contact with corrosive or toxic aterials
- worn whenever there is a possibility of injury to your hands, e.g. when lifting loads manually
- cleaned after use
- replaced periodically depending upon use and type of material handled

7. **Safety shoes**: On a shop floor there is always a possibility of heavy objects falling on your feet or sharp objects puncturing your foot. Hazardous liquids such as chemicals can spill into your shoes and boots. Heavy machinery, equipment, and other objects can roll over your feet often resulting in broken or crushed bones.

8. **Disposable shoe wraps**: These wraps can be worn on safety shoes during manual sun drying to avoid any contamination in the treated products.

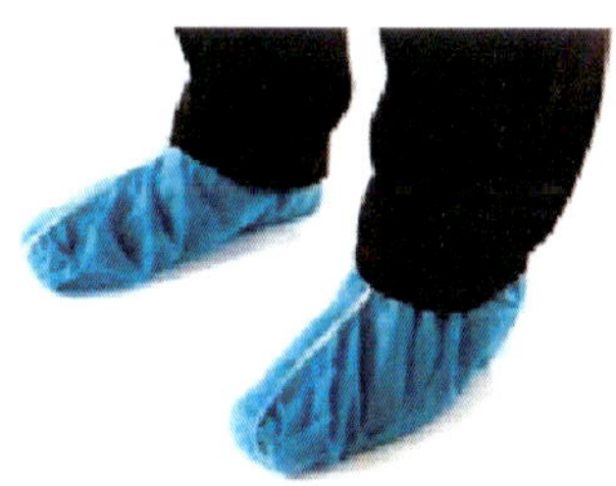

9. **Bird and rat proofing**: This is one of the important aspects of seed storages and whenever seed is kept for shop floor activities like seed processing and treatment. Birds and rats can cause considerable damage to the valuable seed material.

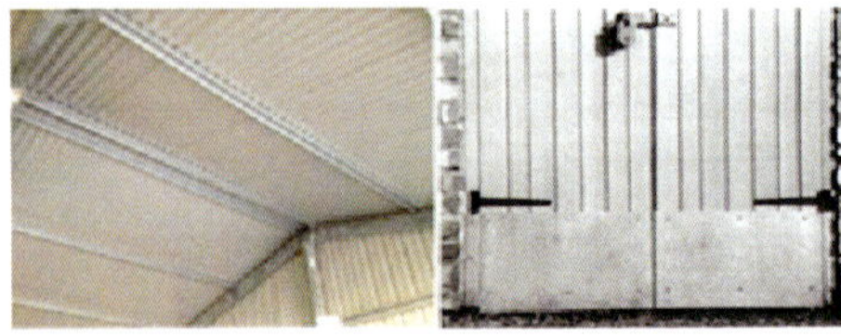

10. **Genetic integrity**: The storage or processing yard should be free from cracks which can hold some seeds may result in mechanical contaminations. The crevises and cracks to be filled with white cement and periodically checked.

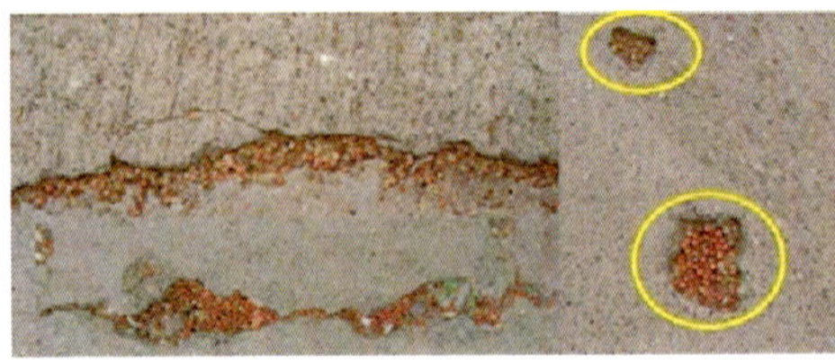

11. **Fire extinguishers**: The storages will hold expensive seed material for very long time. It is better to be equipped with approved extinguishers. a monitored fire detection system should be installed in all treating and storage areas, when and where available. Walls of storage area shall meet appropriate fire and building codes. In general A,B,E and F type extinguishers were used in seed storages. Also sand is standard extinguisher.

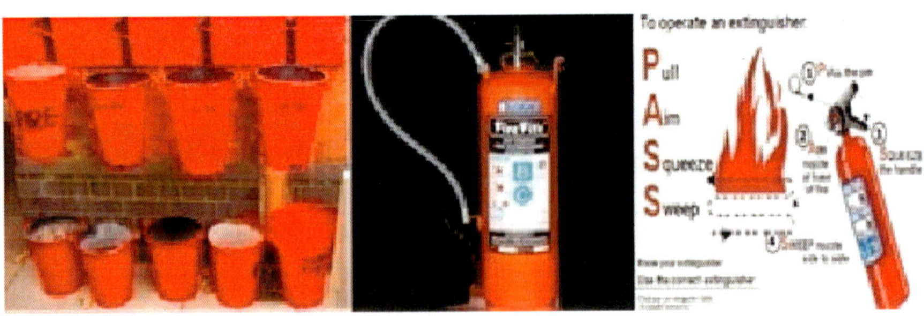


## When selecting and using PPE

- Choose products which are CE marked in accordance with the Personal Protective Equipment Regulations.
- Choose equipment that suits the user – consider the size, fit and weight of the PPE. If the users help choose it, they will be more likely to use it

- If more than one item of PPE is worn at the same time, make sure they can be used together, eg wearing safety glasses may disturb the seal of a respirator, causing air leaks
- Instruct and train people how to use it, eg train people to remove gloves without contaminating their skin. Tell them why it is needed, when to use it and what its limitations are
- Never allow exemptions from wearing PPE for those jobs that 'only take a few minutes'
- Check with your supplier on what PPE is appropriate – explain the job to them
- If in doubt, seek further advice from a specialist adviser

## Maintenance

PPE must be properly looked after and stored when not in use, eg in a dry, clean cupboard. If it is reusable it must be cleaned and kept in good condition.

Think about:

- using the right replacement parts which match the original, eg respirator filters
- keeping replacement PPE available
- having a supply of appropriate disposable suits which are useful for dirty jobs where laundry costs are high, eg for visitors who need protective clothing

**Activity:** Do a ESH and PPE audit for our seed processing unit

*Refer Worksheet-3 for safety audit*

# Practical – 26

# Application of Hagberg Falling Number Method in Predicting High Moisture Damage and Storage of Seeds in Wheat and Paddy

The "falling number" is a measurement used by plant breeders and grain quality specialists to determine the soundness of wheat, which is primarily influenced by the degree of sprouting in a grain sample. Hagberg Falling Number is the full name of this international standard of sprout measurement. The term "falling numbers" refers to the amount of time it takes for the Hagberg steel ball in this sprout analysis tool to fall through a flour/water slurry that has been heated to release the starch from water.

This method can be used for correlating the high moisture damage during storage of seeds. This method can be used for cereals like wheat and paddy. This method work on assuming that when seed moisture content is higher than the safe levels then there will be start of the germination process. If this process commences then the seeds cannot be stored for very long durations because of spoilage of seed quality due to higher amylase activity.

The Falling number test does not directly measure amylase enzyme activity, but measures changes in the physical properties of the starch portion of the wheat kernel caused by these enzymes during the test.

## Procedure

***Step 1:*** Take different seed samples with different moisture content. Make a flour out of it and add water to make a fine slurry

***Step 2***: Stirring rods are inserted into the test tubes, which are placed in the boiling water bath (212° F) in the falling number apparatus. They are automatically stirred for 60 seconds, causing the starch-water slurry to thicken due to starch swelling or "gelatinization." When starch granules become gelatinized, they are increasingly susceptible to degradation by amylase enzymes, which increases the rate of thinning in the slurry.

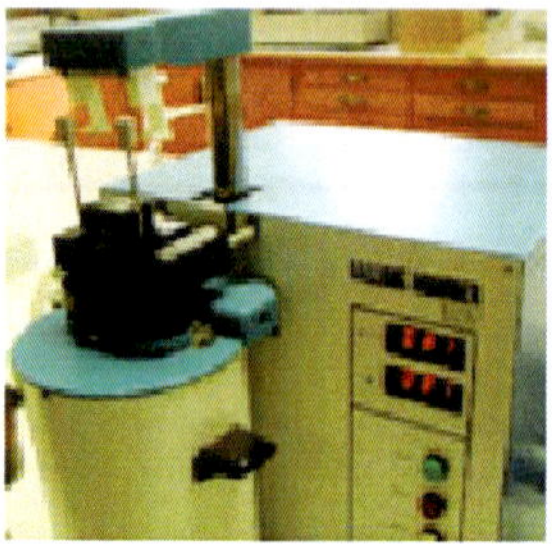

***Step 3***: After mixing and heating, the stirrers are released at the top of the slurry. The falling number apparatus records the time for the stirrer to fall through the slurry. The rate at which the stirrer falls is directly related to the amount of starch degradation by amylase enzymes. A falling number value of 300 seconds or longer indicates seed is better for longer storage. If it fall below 200 then seed can't be stored for longer periods

## Observations

Take seeds with higher moisture content and do the HFN test. Then store for particular time period and then test for germination and vigour. Correlate the HFN values and Germination %.

**Table 1:** Correlation of Moisture % and HFN test

| SMC % | HFN No |
|---|---|
| 8% | |
| 9% | |
| 10% | |
| 11% | |
| 12% | |
| 13% | |
| 14% | |
| 15% | |
| 16% | |

**Table 2:** Correlation of HFN test and Germination % to seed storage

| Storage (M) | HFN Values | Germ% | SVI |
|---|---|---|---|
| 3 | <200 | | |
| | 200-300 | | |
| | >300 | | |
| 6 | <200 | | |
| | 200-300 | | |
| | >300 | | |
| 9 | <200 | | |
| | 200-300 | | |
| | >300 | | |
| 12 | <200 | | |
| | 200-300 | | |
| | >300 | | |

## Study Questions

List out various methods of predicting vigour and viability

# Practical – 27

# Estimation of Post Harvest Seed Losses

There are stages in seed supply chain where seed losses happen at harvesting and post harvest stage. These losses may occur due to higher moisture at harvest, due to harvesting methods, processing methods, at seed storage, during transport etc. Estimation of these losses is an important function of seed inventory management. Estimating seed losses help to improve processing and seed distribution efficiency. These simple statistical tools will help to measure quantitative post harvest seed losses. Seed testing can be used for measuring qualitative post harvest seed losses.

1. **Seed loss due to higher moisture content**: When the seed grower surrenders the seed produce to processing unit, moisture is tested. If the moisture doesn't conforms to the optimum level and if no mechanical or shade drying facilities available or seed growers not ready to dry the material then estimating seed loss due to higher moisture is necessary by the following formula (modified from Boxall (1986).

$$\text{Drywt.of seed} = \frac{\text{wt.of seed x (100-m.c. of seed)}}{100}$$

2. **Seed loss due to insect infestation**: Whenever the seed is stored in un protective environment there will be insect infestation. Here damaged and undamaged grains in a sample of 100-1000 grains are counted and weighted.To estimate seed loss due to insect infestation the following formula can be used (modified from Harris and Lindblad 1978).

$$\%\ \text{weight loss} = \frac{(\text{Und})-(\text{DNu})}{\text{U}(\text{Nd}+\text{Nu})} \times 100$$

Where, U = weight of undamaged seeds,

Nu = number of undamaged seeds,

D = weight of damaged seeds,

Nd = number of damaged seeds.

3. **Seed loss estimation in seed storage**: To estimate the seed loss during storage due to all factors like insects, pests, pathogens and general decaying this formula can be used (modified from Compton, Floyd, Ofosu and Agbo 1998)

Percentage weight loss =

$$\frac{T(Wd+Wu)Wu+FW(NdWu-NuWd)}{TND(Wd+Wu)Wu+FW(Nd+Nu)Wu} \times 100$$

Where, TND = total number of destroyed and missing seeds,

FW = the final weight after the eight-step process,

Nd = number damaged seeds in subsample,

Nu = number undamaged seeds in subsample,

Wd = weight damaged seeds in subsample,

Wu = weight undamaged seeds in subsample.

4. **TSW method of estimating seed loss**: Here the seed loss is estimated based on TSW (thousand seed weight) (modified from Boxall 1986). At the beginning a standard sample TSW (dry weight basis) is calculated and kept as reference. Later the TSW (dry weight basis) of the submitted sample is calculated and seed loss by the formula:

$$\text{Seed loss} = \frac{\text{Initial TGM} - \text{Sample TGM}}{\text{Initial TGM}} \times 100$$

TSW on dry basis can be calculated as follows:

$$MD = \frac{10m\,(100-H)}{N}$$

Where,

MD = TGM (dry basis)

m = mass (weight) of seeds in the sample

N = number of seeds in sample

H = moisture content in sample

## References

Boxall, R.A. 1986. A critical review of the methodology for assessing farm level grain losses after harvest. Report of the Tropical Products Institute, G191, Greenwich, United Kingdom.

Compton, J.A.F., Floyd, S., Ofosu, A., & Agbo, B. 1998. The Modified Count and Weigh Method: An Improved Procedure for Assessing Weight Loss in Stored Maize Cobs. *Journal of Stored Products Research*, 34(4): 277-285.

Harris, K.L. & Lindblad, C.J. 1978. Post harvest grain loss assessment methods. A manual of methods for the evaluation of post harvest losses". The league for International Food Education, The Tropical Products Institute (England), FAO, Group for Assistance on Systems Relating to Grain After- Harvest.

## Study Questions

List out various types of post harvest seed deterioration

# Practical – 28

# Seed Sampling at Processing Plant

To assess the quality of the seed is necessary before it is shipped for packaging and made available to the farming community. To assess certain routine seed quality parameters which are statutory for labelling seed sampling and testing are very critical. Seed testing is not only a mandatory trade requirement but also decides the market potential of a particular product in the market and safeguard the brand image of the company. Seed sampling for labelling and other trade requirements is done at processing plant. The reliability of seed testing is depends on the accuracy with which the sample represents the lot and precision of the laboratory test. In many cases variations in the test results are due to variation in the sampling. Hence, seed sampling is one of the important activities of the shop floor or processing unit.

**Sampling**: A sample is obtained from a seed lot by taking small portions at random from different positions and combining them.

**Seed lot**: It is specified quality of seed, physically identifiable, in respect of which a seed test certificate can be issued.

**Primary sample**: Small portions taken from one point in the lot

**Composite sample**: Is formed by combining and mixing all primary samples taken from the lot.

**Submitted sample**: Sample submitted to seed testing lab. The size of the submitted sample is specified in seed testing rules (***refer Worksheet-5: Size of the submitted samples***)

**Working sample**: Is a sub sample taken from submitted sample in the laboratory, on which one of the quality test is made.

## Procedure for seed sampling

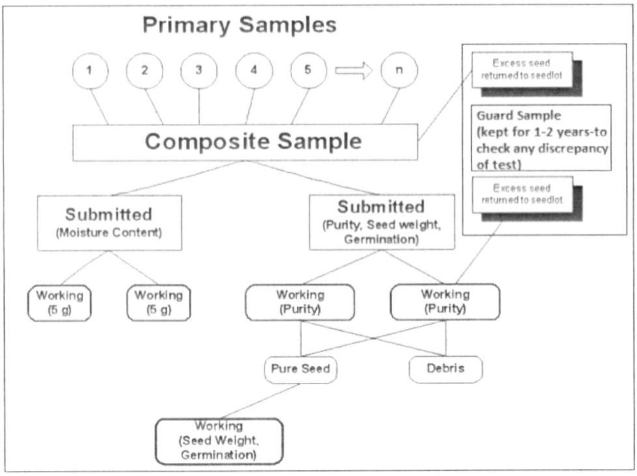


## Sampling intensity

### A: For seed lots in bags

| | |
|---|---|
| up to 5 containers | Sample each container -but never < 5 Primary Sample (PS) |
| 6-30 | Sample atleast one in every 3 containers- but never > than 5 Primary Sample |
| 31-400 | Sample atleast one in every 5 containers- but never < 10 Primary Sample Sample |
| 401 and more | Sample atleast one in every 7 containers- but never < 80 Primary Sample |

### B: In small containers

When the seed is in small containers such as tins, cartons or packets a 100 kg weight is taken as the basic unit and small containers are combined to form sampling units not exceeding this weight e.g. 20 containers of 5 kg each. For sampling purpose each unit is regarded as one container.

### C: For seeds in bulk

| | |
|---|---|
| Up to - 500 kg | Atleast 5 Primary samples |
| 501 - 3000 Kg | One primary sample for each 300 kg but not less than 5 primary samples |
| 3001-20,000 Kg | One primary sample for each 500 kg but not less than 10 primary samples |
| 20,001 and above | One primary sample for each 700 kg but not less than 40 primary samples |

**Sampling equipments:** for seed sampling different sized tiers are used.

**Sleeve or stick type bag trier**: The double sleeve trier consists of a hollow tube with a solid pointed end and a close fitting inner tube such that the product cannot slip between the two sleeves. The inner tube may be with or without partitions (fixed or removable plugs) between the slots. Partitioned double sleeve triers may be used horizontally or vertically. Double sleeved triers without partitions may only be used vertically when sampling sacked grain weighing greater than 100 kilograms. Vertical sampling of sacked grain 100 kilograms or less must be undertaken with the partitions in place.

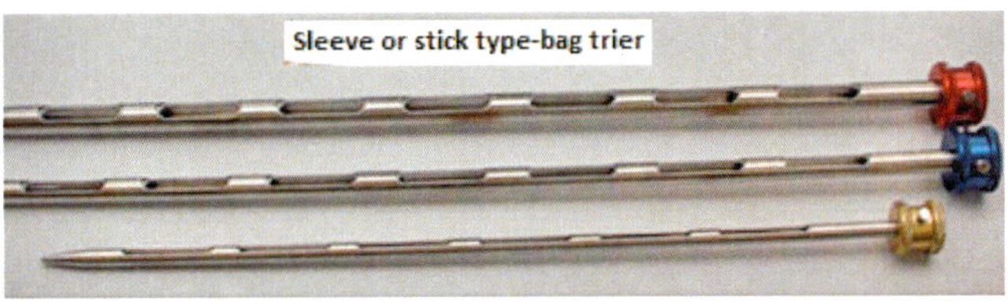


**Nobbe triers**: The Nobbe trier is a pointed tube with an oval opening near the pointed end. It is relatively compact, making it easy to transport. The risk of contamination is low as the trier is easy to keep clean. A Nobbe trier is suitable for sampling free-flowing product in bags, but only where the trier can reach to the centre of the bag. It may only be used horizontally and its use is limited to penetrable containers

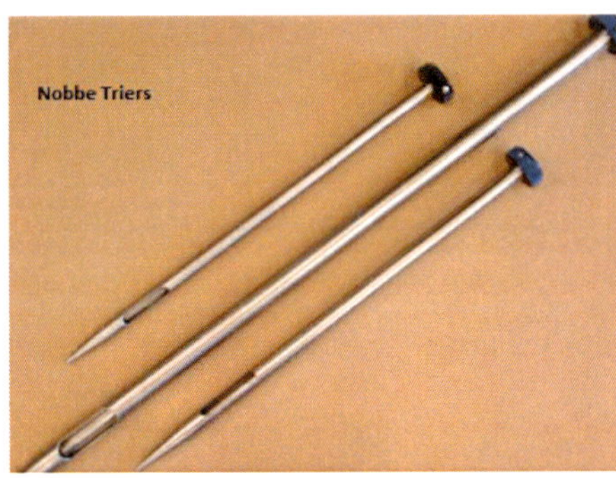


**Format for sending samples**: This format must be used after sampling to send samples to the STL.

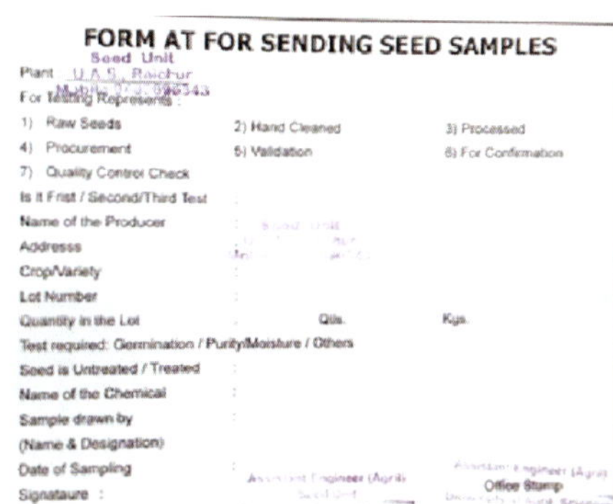

**FORM AT FOR SENDING SEED SAMPLES**

Plant U.A.S., Raichur

For Testing Represents :

1) Raw Seeds 2) Hand Cleaned 3) Processed

4) Procurement 5) Validation 6) For Confirmation

7) Quality Control Check

Is it Frist / Second/Third Test :

Name of the Producer :

Addresss :

Crop/Variety :

Lot Number :

Quantity in the Lot : Qtls. Kgs.

Test required: Germination / Purity/Moisture / Others

Seed is Untreated / Treated :

Name of the Chemical :

Sample drawn by :

(Name & Designation)

Date of Sampling :

Signataure : Office Stamp

## Practical – 29

# Management of Stacking and the Problem of "Stackburn"

The management of storage in open and closed storage is critical before start of any seed processing activities. When seed lots reach the shop floor, they need to be stored properly. Unlike commercial produce seed lots cannot be stored in bulk. These need to be stacked lot wise and there is problem of "stackburn"-resulting from successive layers in building a bag stack. In crops like soyabean-"stackburn" can have detrimental effect on seed quality. The quality parameters like germination and viability can get lost instantly if not properly stacked on the pallets.

### Management of stacking

1. Building suitable channels and chimneys which will provide ventilation of the stack centre (Fig.1).
2. The layers of bag stacks will be modified by providing proper channels running from the bottom to the top of the stack while maintaining its stability (Fig.2).

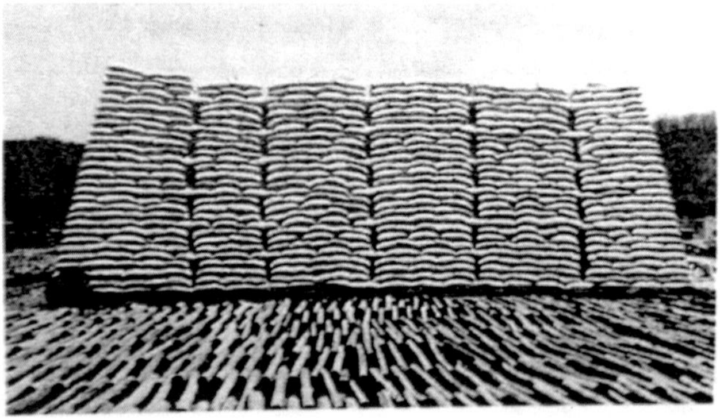

**Fig.1:** Channels for passive ventilation and dunnage in foreground

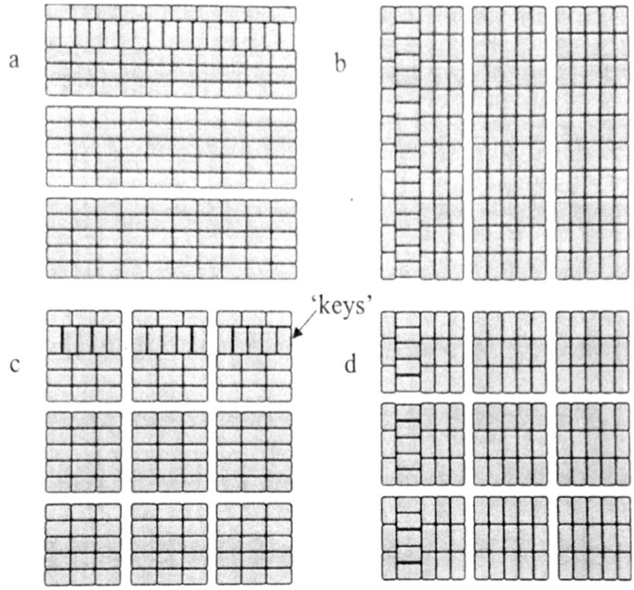


**Fig. 2:** Successive layers in building a bag stack with channels to limit "stackburn"
a) Stacking with no channels
b) Channels right angles to the layers below
c) &d Channels in both the directions

## Points to remember in stacking management

1. Stacks of seed bags were covered with a tarpaulin in wet season and during dry season no need of covering. During wet season or rainy season these covers need to be opened at least once a week which allows hot air out of the stack.

2. A layer of sawdust or paddy straw or ground nut straw will be placed between tarpaulins and bags to absorb moisture generation due to condensation.
3. Using plastic or rubber pallets for stacking is first step in any stacking. Tall wooden pallets will provide better ventilation for the bottom bags
4. The size of the stacks should be reduced to increase the surface area to volume ratio. This has positive effect in enhancing passive ventilation (cooling) and reducing grain discoloration. This will increase the efficiency of fumigation.

## Observations

1. Take samples from the stacks which have no channels
2. Take samples from stacks from tall stacks and stacks with no pallets
3. Take samples from stacks which have channels one and both the directions
4. Check the germination % and see for the variations in seed quality

| Sl No | Stack Type | GE % (R-1) | GE % (R-2) | GE % (R-3) |
|---|---|---|---|---|
| 1 | No-channels | | | |
| 2 | Tall stacks | | | |
| 3 | No-pallets | | | |
| 4 | Channels-one direction | | | |
| 5 | Channels-both directions | | | |

## Study Questions

List out mechanism of seed deterioration in storage

# Practical – 30

# Deciding Suitability of Paddy Seed for Drying and Storage

Storage of paddy seed lots after harvesting is one of the crucial aspects of processing shopfloor. Paddy due to increased mechanised harvesting may loose vital seed quality parameters. Since seed growers cannot buy a combined harvester themselves-they resort to hiring of this machine. Sometimes in the hurry of completing harvesting they harvest the crop at un-safe moisture content. This un-safe moisture content may go up to 50-60% also. Harvesting of seed crop with high moisture has adverse effect on storability. Ensuring safe seed moisture is very critical especially during rabi-summer seed production cycle. Most of the growers depend on sun drying for reducing the seed moisture. But paddy should not be dried >3% in one cycle. Follow tempering and convection cooling after every cycle and repeat the cycle till seed attains safe moisture content (14%).

## Exp-1: Drying of paddy seed having high moisture

Target safe moisture: 14%

| Seed MC% | Moisture % per step drying | No of steps |
|---|---|---|
| 30 | 3 | 5 |
| 25 | 3 | 4 |
| 20 | 3 | 2 |
| 18 | 3 | 1 |
| 16 | 3 | 1 |

**Problem**: Though the final seed produce may have safe moisture content but will have less storability. To check suitability of paddy seed lots for storability is the aim of this experiment. The suitability for storability is calculated by the 'evaluation value'

## Evaluation Value equations for paddy:

Y (Indica paddy) = 65.7+ (0.07 X germ rate) - (0.25 X free fatty acid value) + (1.7 X viscosity)

Y (Japonica paddy) = 83.3+ (0.005 X germ rate) - (0.25 X free fatty acid value) + (1.7 X viscosity)

Seed is considered UNSUITABLE for storage if the evaluation value falls below 70 or if FFA (mg KOH/100 g dry basis) value is above 25 mg KOH/100 g or viscosity less than 4.5 $mm^2s^{-1}$ (Table 1)

**Table 1:** Quality control classification for storability of paddy seed

| **a) Indica type** | | | |
|---|---|---|---|
| Parameters | Suitable | Un-suitable | Aged |
| Evaluation Value | ≥ 70 | < 70 | - |
| Fatty Acid Value | ≤25 | 25-32 | > 32 |
| **b) Japonica Type** | | | |
| Parameters | Suitable | Un-suitable | Aged |
| Evaluation Value | ≥ 70 | < 70 | - |
| Fatty Acid Value | ≤ 25 | 25-32 | > 32 |
| Viscosity | ≥ 10 | 10-4 | < 4 |

# Worksheets

## Worksheet-1: Measurement of physical properties of seeds

Type of seeds: gum guar

| Sl no | Physical Properties | Methods |
|---|---|---|
| 1. | Lenght -cm | Using vernier calipers (Mitutoyo Vernier Calipers: 530-312, 0.01 mm) |
| 2. | Width-cm | |
| 3. | Thickness-cm | |
| 4. | Diameter-cm | |
| 5. | Arithmatic Diameter-cm | calculated using principal dimensions (Bahnasawy, 2007) |
| 6. | Geometric diameter-cm | calculated using principal dimensions (Bahnasawy, 2007) $Dg(mm)=(l.b.t)^{1/3}$ |
| 7. | Volume-$cm^3$ | Unit volume, Vu (cm3) was calculated by assuming that seeds and kernels are similar to triaxial ellipsoid (i.e., l > b > t) (Mohsenin, 1986) $V_{u\,(cm^3)} = \frac{\left[\frac{4}{3}\pi.(l.b.t)\right]}{1000}$ |
| 8. | Sphericity-Sϕ | calculated based on the isoperimetric properties of the sphere (Mohsenin, 1986) $S\varphi = \frac{Dg}{i}$ |
| 9. | Bulk density-ρb ($kg/m^3$) | Bulk density, ρb (kg/m3), of seeds and kernels were determined according to Singh and Goswami (1996): |
| 10. | True density- ρs ($kg/m^3$) | True density or solid density, ρs (kg/m3) was determined by toluene displacement method (Singh and Goswami, 1996). |

| | | |
|---|---|---|
| 11. | Porosity- ε(%) | Porosity, ε(%) was calculated using the values of bulk and solid density (Mohsenin, 1986): |
| 12. | Angle of repose- θ | The angle of repose of seeds and kernels was measured by emptying method in bottomless cylinder (diameter, 5 cm; height, 10 cm) (Taser *et al.*, 2005; Garnayak *et al.*, 2008): |
| 13. | Surface area- *S*a (mm2) | Surface area, Sa (mm2) of seeds and kernels was calculated using the formula proposed by McCabe et al. (1986): |
| 14. | Specific surface area- *S*s (cm2/cm3) | Specific surface area, Ss (cm2/cm3) was calculated (Sirisomboon et al., 2007): |
| 15. | Coefficient of static friction (μ) | Coefficient of static friction (i) was determined on five different surfaces (wood, plywood, stainless steel, galvanized iron and glass), which are commonly used for the processing and handling of pulses and grains (Balasubramanian, 2001): |
| 16. | Hydration capacity (*H*c) | hydration capacity (Hc) and hydration index (Hi) were determined according to the method proposed by (Adebowale et al., 2005): |
| 17. | Hydration index (Hi) | |
| 18. | Swelling capacity (Sc) | To determine swelling capacity (Sc) and Swelling index (Si) were determined (Adebowale et al., 2005): |
| 19. | Swelling index (Si) | Swelling index, Si is the ratio of swelling capacity of seed or kernel to that of volume |
| 20. | Rupture force (Rf) | To determine the rupture force (Rf) the seed was placed on a stationary platform along horizontal and vertical directions and pressed with a moving platform. Rupture force (Rf) and energy for rupture (Er) of gum guar was determined using texture analyzer (TA.XT plus, England). |
| 21. | Colour | Colour value L, a, b and hue angle were measured using Hunter Lab Colorimeter (Kimura et al., 1993) |

## Worksheet-2: Measuring mechanical injury in soya bean and preparation of 1% Sodium hypochlorite solution

Why this test ?: In case of soya bean the seed coat, or testa, of these seeds is very thin and easily cracked. In addition, the plumule and radicle, parts of the seed embryo, are just under the seed coat and can easily be bruised by rough handling. Because they are dicotyledons, these plants can easily be broken in two.

Materials required:

- Soybean seeds
- Freezer bags (quart size)
- Laboratory beakers - (1-1000 ml and 2-250 ml)
- Household bleach
- Plastic pan or paper towels

Preperation of 1% sodium hypochlorite solution: Mix a 1% sodium hypochlorite solution in a 1000 ml beaker using household bleach and water. Household bleach contains about 5% sodium hypochlorite so you can mix 1 part bleach with 4 parts water to form the 1% solution.

Alternatives to bleach: Hot water, and even coffee, will work in the place of bleach, but may take longer

Procedure:

1. Weigh out two, 40-50 g samples of soybean seed.
2. Collect the seed samples from seed grader and Place one of the samples in a freezer bag. Also you can induce damage by dropping the bag on a hard surface 10-15 times to damage the seeds in case you don't have seed grader provided about 5% of the seeds should be broken in half when you are finished.
3. Place the seed from the freezer bag that you dropped on a hard surface in a 250 ml beaker containing 1% sodium hypochlorite solution. Place the sample of undamaged soybean seeds in another 250 ml beaker of 1% sodium hypochlorite solution.
4. To calculate exact percentages take 100 seeds in 4 replications and calculate percentage of mechanical damage
5. Allow the beans to soak for a few minutes and then observe the cracked seeds in each lot. The seed coat of the cracked soybeans will be swollen and separated from the cotyledons.
6. Check again in 15 minutes and record the percent of cracked seed coats. To make the observation of the seeds easier, strain the sodium hypochlorite from the beaker, rinse the seeds several times with water to get rid of any residual bleach, and gently spread the seed out in a single layer in a plastic pan or on a paper towel.

## Worksheet-3: Internal Audit Form – PPES and ESH

Name of The Seed Processing Unit With Code:

Date of Audit:
Audit Completed By:

| Sl no | ITEM | Yes / No | If YES score (1 to 5: 1 is Poor and 5 Very Good); If NO give reason for not implementation |
|---|---|---|---|
| **A. Personal Protective Equipment (PPE):** | | | |
| 1 | Whether all these PPEs are available: | | |
| 2 | Helmets | | |
| 3 | Safety Glasses / Goggles | | |
| 4 | Aprons | | |
| 5 | Ear Protection (ear-plugs/ear-muffs) | | |
| 6 | Masks | | |
| 7 | Gloves | | |
| 8 | Safety shoes | | |
| 9 | Shoe wraps and others | | |
| 10 | Has adequate storage been provided for PPE? | | |
| 11 | Have checks been made to ensure that PPE fits the wearer properly | | |
| 12 | Has PPE been regularly visually inspected and records of the inspection kept? | | |
| 13 | Has PPE been repaired when necessary? | | |
| 14 | Has PPE been replaced when it is worn out/defective? | | |
| 15 | Are staff aware of when to wear the appropriate PPE and why? | | |
| 16 | Do staff fully understand the risks they are protecting themselves against? | | |
| 17 | Where gloves are used, have the appropriate ones been selected for the substances in use? | | |
| **B. Environment, Safety and Health Monitoring** | | | |
| 18 | Whether the shop floor is well floored-cracks and crevices are filled up? | | |
| 19 | Bird and rat proofing is done or not? | | |
| 20 | Whether the machinery cleaned properly before start of processing and treatment? | | |
| 21 | Whether the processed material is properly stored? | | |
| 22 | Whether recommended stacking is being followed? | | |
| 23 | Whether the pallets are used? | | |
| 24 | Whether the stack cards are used? | | |

25 Whether the seed fumigation is done frequently?
26 Whether sanitation is followed (like sprays on bags)?
27 Whether the rejected seed is disposed timely?
28 Whether fire extinguishers are in place?
29 What is the due date for servicing (fire extinguishers)?
30 Whether all the machines are properly serviced?
31 Whether operating manuals for the machines are available?
32 Whether the spares are available?
33 Whether any breakdown registers are maintained?
34 Whether the machines are operating to the fullest potential?
35 Whether seed treatment chemicals are stored properly?
36 Whether moisture proof bags are used?
37 Whether the gunny bags are in good conditions?
38 Whether this facility has rejected seed storage godown?
39 Whether this plant issue any seed receipt to growers? (after receiving the material from field)
40 Has processing register is maintained?
41 Whether processing reports are well maintained?
42 Whether follow STL reports regularly?
43 Whether the seed moisture register is maintained? (both before processing and before packing)
44 Whether the seed receiving area has standard balance?
45 Whether this plant check seed moisture on arrival?
46 Whether the cold storages are maintaining the required RH and TEMPERATURE?
47 Whether the sampling is done as per the guidelines of KSSCA?
48 Whether there is good hygiene facilities are available? (wash rooms and toilets)
49 Whether SOPs for all activities (standard operating procedures) are available?
50 Whether training facilities and information on available seeds are readily available? (brochures, seed samples etc)

Any other points you wish to highlight?

Action Points Date Action Taken

(What action is to be taken as a result of completing this monitoring- checklist?)

Date Completed: ...................................... Signed: ....................................

**Worksheet -4: Various seed quality parameters and quantification methods**

| Sl no | Test | Test Developed By | Significance and Formula |
|---|---|---|---|
| 1 | Volume based Imbibition Time (Hr) | Leopold (1983) | Time taken for 2mm emergence of primary root from the seed coat. $Kt=-\log_e (a_{max}-a)/a_{max}$ Where $a$ is the volume at time $t$, $a_{max}$ is the volume at full hydration(24 h) and $k$ is swelling co-efficient |
| 2 | Germination (%) | Nichols *et al.*, (2009) | Seeds were considered to have germinated when radicles reached at 5 mm length Germination percentage = (No. of germinated seeds / Total No. of seeds sown) X 100 |
| 3 | Rate of Germination | Agrawal (2004) | Rate of Germination $\Sigma (ni / Di)$ $ni$ is the number of germinated seed on day $i$ and $Di$ is the number of days after beginning the experiment. |
| 4 | Mean Germination Time | Ellis and Roberts (1981) | MGT= $\Sigma(Dn)/\Sigma n$ Where $n$ is the number of germinated seeds or emerged seedlings on day D and D is the total number of days counted from the beginning of germination. |
| 5 | Mean Daily Germination | Czabator (1962) | MDG = Total number of germinated seeds/ Total number of days |
| 6 | Peak Value | Czabator (1962) | PV = Highest seed germinated/ Number of days |
| 7 | Germination Value | Czabator (1962) | GV= Peak value X Mean Daily Germination |
| 8 | Emergence Rate Index | Adetimirin (2008) | ERI estimates the total number of days for 100% emergence to occur |
| 9 | Shoot Length (cm) | Agrawal (2004) | Ten normal seedlings of each replication were taken and shoot length was measured in centimeters from collar region to the tip of the shoot and average of ten seedlings werecomputed. |
| 10 | Root Length (cm) | Agrawal (2004) | Ten normal seedlings from each of the replication from germination test were removed carefully and used for measuring root length. Root length was measured in centimeters from the collar region to the tip of main root with the help of scale. The average of ten seedling roots was computed and expressed in centimeters. |
| 11 | Seedling Vigour Index | Abdul Baki and Anderson (1973) | The seedling vigor index (SVI) = (Seedling length (cm) X Germinated percentage) / 100 |

| | | | |
|---|---|---|---|
| 12 | Seedling Dry Weight (mg) | Abdul Baki and Anderson (1973) | Dry weight of seedlings were recorded after drying them in the hot air oven at 700C for 48 hours and they are cooled by keeping in a desiccator for one hour and the average dry weight of the seedlings was expressed in milligrams. |
| 13 | Relative Dry Weight | Czabator (1962) | Increase of dry weight of seedlings over the control |
| 14 | Seed Viability (%) | Perry (1987) | Seed viability was determined by a tetrazolium test using 400 seeds, placed in three separate replications of 100 seeds |
| 15 | Usable Transplants (%) | Hartley *et al.*,(2001) | Percentage of healthy, well grown, saleable seedlings or transplants obtained per gram of seed at 10–14 days after sowing. |

## Worksheet-5: Size of the submitted samples

| Crop | Size of seed lot (Kg) | Size of submitted sample (g) | Size of working Sample for purityanalysis (g) | Sample count of other species(g) |
|---|---|---|---|---|
| Paddy | 20,000 | 400 | 40 | 400 |
| Wheat | 20,000 | 1000 | 120 | 1000 |
| Maize | 40,000 | 1000 | 900 | 1000 |
| Sorghum | 10,000 | 900 | 90 | 900 |
| Bajra | 10,000 | 150 | 15 | 150 |
| Redgram | 20,000 | 1000 | 300 | 1000 |
| Greengram | 20,000 | 1000 | 120 | 1000 |
| Blackgram | 20,000 | 1000 | 150 | 1000 |
| Bengalgram | 20,000 | 1000 | 1000 | 1000 |
| Cowpea | 20,000 | 1000 | 400 | 1000 |
| Soybean | 20,000 | 1000 | 500 | 1000 |
| Groundnut(pods) | 20,000 | 1000 | 1000 | 1000 |
| Groundnut(Kernels) | 20,000 | 1000 | 600 | 1000 |
| Gingelly | 10,000 | 70 | 7 | 70 |
| Sunflower(variety) | 20,000 | 1000 | 250 | 1000 |
| Sunflower(Hybrid) | 20,000 | 1000 | 125 | 250 |
| Cotton linted(variety) | 20,000 | 1000 | 350 | 1000 |
| Cotton delinted(variety) | 20,000 | 350 | 35 | 350 |
| Cotton linted(hybrid) | 20,000 | 350 | 35 | 350 |
| Cotton delinted (hybrid) | 20,000 | 250 | 25 | 250 |
| Brinjal | 10,000 | 150 | 15 | 150 |
| Chillies | 10,000 | 150 | 15 | 150 |
| Bhendi | 10,000 | 150 | 15 | 150 |
| Tomato(variety) | 10,000 | 70 | 7 | 70 |
| Tomato(hybrid) | 10,000 | 7 | 7 | 7 |
| Cabbage | 10,000 | 100 | 10 | 100 |
| Cauliflower | 10,000 | 100 | 10 | 100 |
| Knolkhol | 10,000 | 100 | 10 | 100 |

(Collect more crops in the space provided)

## About the Authors

**Rakesh C. Mathad**, a seed technologist by profession having more than 14 years of experience in seed industry and academics. He is presently working as Assistant Professor-Seed Science and Technology at Seed Unit, University of Agricultural Sciences, Raichur, Karnataka. He is an expert in seed production and production research. He has published more than 15 research articles in many national and international journals of repute. He has authored and co-authored four books and visited countries like China, Israel, Turkey and Ethiopia to attend international conferences and short courses. He has developed molecular genetic purity tests for many pulses and also have patents on devises used in hybrid seed production. His area of interest includes seed physiology and storage, seed production, seed conditioning, seed proteomics, molecular aspects of seed testing and nano technology.

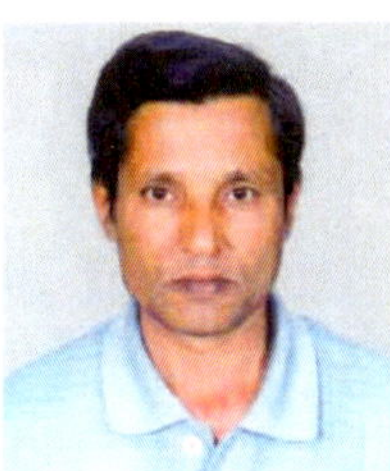

**A.S. Channaveeraswamy,** currently works as Farm Superintendent ARS, Annigeri, UAS, Dharwad, Karnataka have more than 25 years of experience in research and academics. His areas of specialization include seed production, seed testing, quality control, seed processing and seed health management. He has published 30 full length papers, 25 abstracts, popular articles and book chapters. He guided 14 MSc and 2 PhD students.

**B. S. Vyakaranahal,** currently works as professor of seed science and technology, Agriculture College, UAS, Dharwad, Karnataka have more than 30 years of experience in academics. His areas of specialization include seed production, storage and testing. He was instrumental in establishing and strengthening department of seed science and technology at UAS, Dharwad. He also contributed towards formulation of course curriculum of UG and PG programmes. He guided 20 MSc and 8 PhD students. He has published 100 full length papers, 80 abstracts, popular articles, technical bulletins and book chapters. He is a Fellow of Indian Society of Seed Science and Technology, New Delhi. He is also recipient of many awards including "Best Reviewer" from ARCC. Karnal.

**V.K. Deshpande** currently working as professor of seed science and technology at Agriculture College, UAS, Dharwad, Karnataka have more than 20 years of experience in academics. His areas of specialization include seed processing, seed quality enhancement and storage. He has published 52 full length papers, 100 abstracts, popular articles and book chapters. He authored many books in Kannada language for the benefit of seed growers and farmers. He has visited USA along with PG students on exposure trip to Cornell University. Ithaca, New York (Part of INTAG Course). He guided 16 MSc and 5 PhD students and recipient of many awards like "*Krishisri*". He is also a Fellow of Indian Society of Seed Science and Technology, New Delhi.